Table of Contents

C.E.	Inventors and Their Inventions	Page
	The Abbasid Caliphate Timeline	5
	Introduction	7
721	The Battle of the Zab / Birth of the Abbasid Caliphate Birth of the Golden Age of Islam	9
721	Abu al-'Abbās 'Abdu'llāh as-Saffāḥ The 1st Caliph	11
722	Jabir Ibn Hayyan Hydrochloric and Sulfuric Acid	13
751	Papermaking	15
754	Abu Ja'far Al-Mansur City of Baghdad / City of Peace	17
766	Habash Al-Hasib Al-Marwazi *First Calculation of Time*	19
770	The House of Wisdom	21
776	Al-Jahiz Natural Selection	25
780	Muhammad Bin-Musa al-Khwarizmi Algebra and Computer Science	29
786	Al Ma'mun Muslim Doctrine	31
800	Fatima Al-Fihri Al Qarawiyyin Mosque and University	33
801	Al Kindi First Muslim Philosopher	35
803	The Banu Musa Brothers "Sons of Moses"	37
809	Hunayn Bin Ishaq Ophthalmology	41
825	Taqi al Din Al-Shammisiyyah Observatory	43
835	Ahmad Bin Tulun Economist	45

836	Thabit Bin Qurra Trepidation of the Equinoxes	49
840	Al-Adli ar-Rumi Chess Manual	51
858	Al-Battani Astronomical Calculations	53
864	Abū Bakr Muhammad Bin Zakariyyā al-Rāzī Differentiated Measles from Smallpox	55
870	Abu Nasir al Farabi Logic, Physics, Music, Metaphysics, and Politics	59
875	Abbas Ibn Firnas The First Human Flying Machine	61
879	Ahmad ibn Fadlan Diplomat, Traveler and Viking Expert	63
882	Saadia Gaon Jewish Philosopher and Theologian	65
893	Yahya Bin Adi Mansur Theorist and Doctor	67
896	Ali Bin Al Husayn Al Masudi Herodotus of the Arabs	69
903	Al Sufi Nebulosity of the Nebula	71
920	Abu'l-Hasan al-Uqlidisi Decimal Fractions / Arabic Numerals	73
932	Ali Abuzar Mari The Fountain Pen	75
933	Al-Hakim Nishapuri Imam of the Muhaddithin	77
935	Abu Qasim Ferdowsi Persian Poet	79
936	Al Zahwari Inhalation Anesthesia	81
940	Abu Al Wafa Buzjani Astronomer	85
944	Mariam Al-Astrolabiya Astronomer	87

946	Muḥammad Bin Aḥmad Al- Muqaddasi Geographer	89
965	Bin al-Haytham The Camera Obscura	91
973	al-Biruni First Anthropologist	97
975	Al-Mu'izz Il Din Allah Al-Azhar University	99
980	Bin-Sina (Avicenna) Physician	101
987	Prince Abd Al-Rahman Grand Mosque of Cordoba	103
988	Ali Bin Ridwan Supernova	107
1058	Abu Hamid Al Ghazali Theologian	109
1080	Abu Bakr Philosopher	111
1091	Bin Zuhr Cardiologist and Nutritionist	113
1100	Muhammad Al Idrisi Cartographer	113
1126	Bin Rushd (Averroes) Polymath	117
1136	Al-Jazari Mechanical Inventions	119
1137	Yusuf ibn Ayyub ibn Shadi Saladin Sultan of Egypt and Syria	125
1149	Fakhr al-Din al-Razi Theologian	127
1154	Sohrevardi Philosopher of Illusionism	129
1191	Melike Mama Hatun Turkish Bath	131

1200	Mo'ayeduddin Urdi Model of Planetary Motion	133
1201	Nasir Al din Al Tusi Rational Science	135
1213	Bin Al-Nafis Physician	137
1236	Yahya Bin Mahmud al-Wasiti Maqamat of Al-Hariri	139
1237	Spinning Wheel	145
--	Madrasa	147
	Bibliography	148

The Abbasid Caliphate Timeline

1st	750-754 CE	al-Saffah
2nd	754-775	al-Mansur
3rd	775-785	al-Mahdi bi'llah
4th	785-786	al Hadi
5th	786-809	Harun al-Rashid
6th	809-813	al-Amin
7th	813-833	al-Ma'mun
8th	833-842	al-Mu'tasim bi'llah
9th	842-847	al-Wathiq bi'llah
10th	847-861	al-Mutawakkil aia'llah
11th	861-862	al-Muntasir bi'llah
12th	862-866	al-Musta In bi'llah
13th	866-869	al-Mu'tazz bi-'llah
14th	869-870	al-Muhtadi bi-'llah
15th	870-892	al-Mu'tamid 'ala 'llah
16th	892-902	al-Mu'tadid bi-'llah
17th	902-908	al-Muktafi bi-'llah
18th	908-929 and 929-932	al-Muqtadir bi-'llah
19th	929 and 932-934	al-Qahir bi-'llah
20th	934-940	al-Radi bi-'llah
21st	940-944	al-Muttaqi li-'llah
22nd	944-946	al-Mustakfi bi-'llah
23rd	946-974	al-Muti li-'llah
24th	974-991	al-Ta'I' -amri 'llah
25th	991-1031	al-Qadir bi-'llah
26th	1031-1075	al-Qa'im bi-amri 'llah
27th	1075-1094	al-Muqtadi bi-amri 'llah
28th	1094-1118	al-Mustazhir bi-'llah
29th	1118-1135	al-Mustarshid bi-'llah
30th	1135-1136	al-Rashid bi-'llah
31st	1136-1160	al-Muqtafi li-'amri 'llah
32nd	1160-1170	al-Mustanjid bi-'llah
33rd	1170-1180	al-Mustadi bi-amri 'llah
34th	1180-1225	al-Nasir li-Din Allah
35th	1225-1226	az-Zahir bi-amri'llah
36th	1226-1242	al-Mustansir bi-'llah
37th	1242-1258	al-Musta'sim bi-'llah

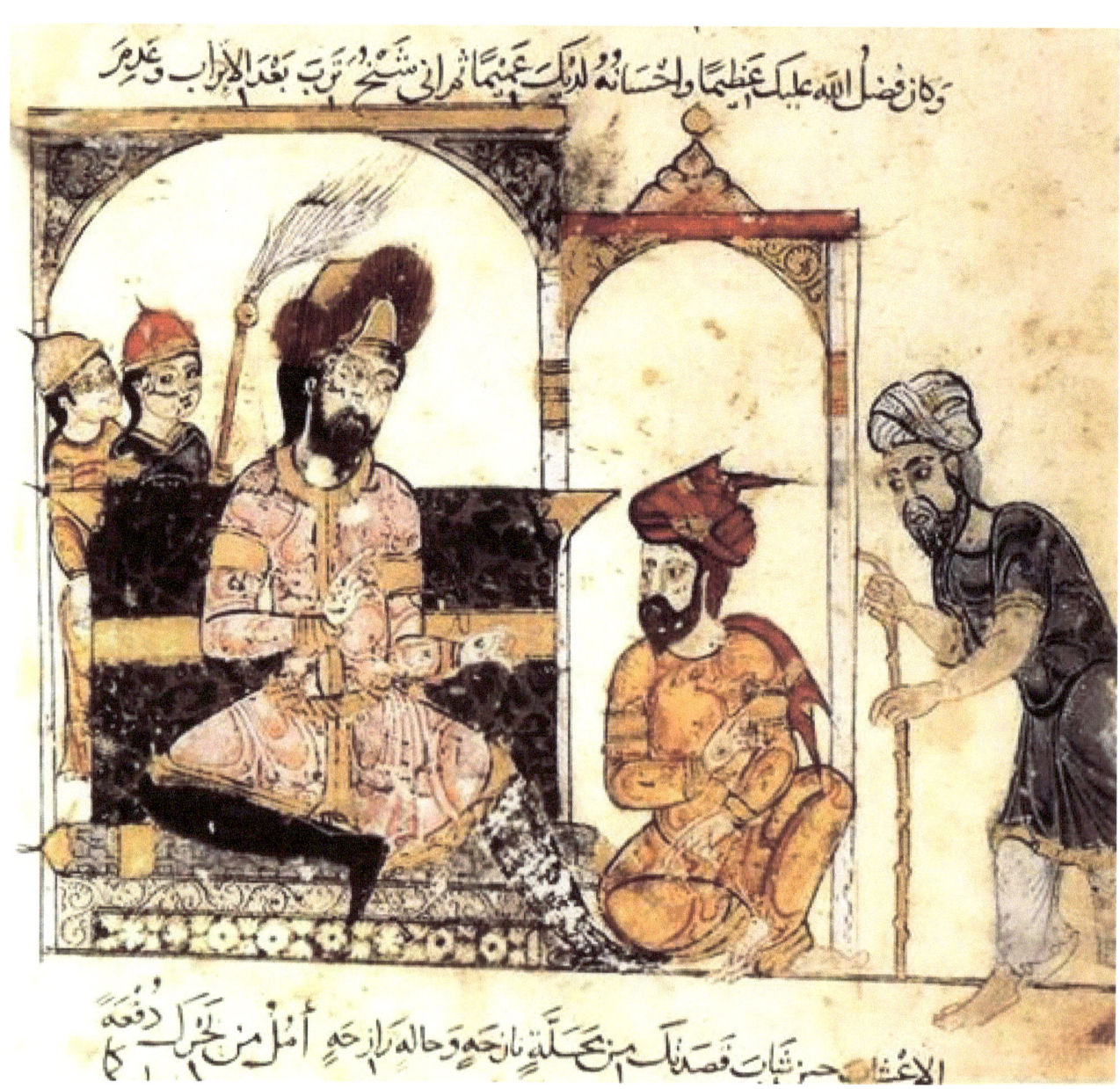

Introduction

Modern history books, and internet research on the most influential astronomers of all time fail to include astronomers from the Golden Age of Islam. Research will provide the following 10 leading astronomers: Claudius Ptolemy (2nd Century CE); Tycho Brahe (1546-1601); Galileo Galilei (1564-1642); Johannes Kepler (1571-1630); William Herschel (1738-1822); Pierre-Simon Laplace (1749-1827); Sir Arthur Eddington (1882-1944); Edwin Hubble (1889-1953); Gerhard Kuiper (1905-1973) and Carl Sagan (1934-1996).

Only one of these 10 astronomers was born prior to 1546. So, what about the astronomers of the Golden Age of Islam? This book will introduce you to many Islamic astronomers who have been denied their historical and timeless significance.

What most students of history will learn for this same period is the Dark Ages, the Crusades, and the Black Plague but not the Golden Age of Islam. When the rest of Europe and Asia were losing all sense of what they had become up until the 8th Century, Islam was relatively new since the founding of the world religion of Islam by the prophet Muhammad ibn Abdullah (570-632 CE).

This collection will introduce readers to the leading minds of the Golden Age of Islam, their great personal achievements, and incredible contributions to both the people of Islam and future generations of Europe, Asia, and the world. It will also give you tidbits of history you can share with your family, friends, and children.

- Steven Lee Douglas

721 CE
Battle of the Zab
Birth of the Abbasid Caliphate
Birth of the Golden Age of Islam

The Battle of the Zab marked the end of the Umayyad Caliphate and the start of the Abbasid Caliphate. The Abbasid's, the 3rd Caliphate since the time of prophet Muhammad, remained in power until the 13th Century. The period of the Abbasid Caliphate is also known as the Golden Age of Islam. The battle was named for the river where the main battle took place. The River Zab is now located in present day Iraq.

The defeated army consisted of Syrian Arabs who marched from Khurasan, located in present day Afghanistan. [i] The Abbasid Caliphate took its name from Muhammad's uncle, Abbas bin Abdul-Muttalib.

The caliphate ruled from their capital Baghdad after defeating the armies of the Umayyad Caliphate at the Battle of the Zab in 750 CE. The first capital of the Caliphate was in Kufa until 762 when the caliph Al-Mansur founded the City of Baghdad. The city was located near the ancient Sasanian capital city of Ctesiphon.

The Abbasid era relied heavily on the Barmakid family who governed the expanding territories of the Caliphate and were credited with increasing the inclusion of non-Arab residents. The Barmakid's were Persian elites who brought Persian customs, which included the patronage of artists and scholars.[ii] The new capital soon became the center of science, culture, philosophy, and inventions, leading to the Golden Age of Islam. During the next two centuries, the Abbasid's ceded their authority over Al-Andalus (Spain) to the Umayyad's (756 CE); Morocco to the Idrisid Dynasty (788 CE); Ifrqiya to the Aghlabids (800 CE) and Egypt to the Isma'ili-Shia Caliphate in 969.[iii] / [iv]

The era of the caliphs ended with the rise of the Iranian Buyids and Seljuq Turks, both of which captured Baghdad respectively in 945 and 1055. The official end of the dynasty came in 1258 when the Mongols led by Hulagu Khan sacked the City of Baghdad.[v]

721 CE
Abu al-ʿAbbās ʿAbduʾllāh as-Saffāḥ
The 1st Caliph

In 750 CE, As-Saffāḥ (721-754), at the age of 29, became the first caliph of the Abbasid Caliphate, one of the longest and most important caliphates (Islamic dynasties) in Islamic history. "As-Saffāḥ", translates as "the Blood-Shedder" for his ruthless tactics he used to instill fear in his enemies.[vi] His time as caliph lasted only four years until his untimely death at 33 years of age.

722 CE

Jabir Ibn Hayyan

Hydrochloric and Sulfuric Acid

Jabir ibn Hayyan was born in the year 722 and died in 815. Jabir is our first inventor of the Golden Age of Islam, known for perfecting chemical processes. These processes included distillation, crystallization, and evaporation. He was also known for discovering hydrochloric and sulfuric acid. Jabir's father Ja'far bin Yahya was originally in the tribe Al-Azd. Yahya was executed shortly after Jabir's birth. Jabir and his family fled to Arabia afterward. He became close to the Barmakid Family (also called the Barmecides) and was allowed to work in secret with their protection. The Barmecides were an influential family that grew in power under the reign of as-Saffāḥ, the 1st Abbasid Caliph. Jabir later fled to Kufa where he died at 93 years old.[vii]

751 CE
Papermaking

Although not invented in Islam, papermaking was especially important during the Golden Age of Islam. Until Muslim forces captured Chinese during the war of 751, little was known of the technical knowledge of modern papermaking. The Chinese captives miraculously were knowledgeable about the Chinese Art of Papermaking and agreed to educate their captors in return for their freedom. Once this knowledge was acquired, factories were built throughout the Muslim countries and papermaking became widely used to publish the Qur'an and other important messages the Caliphate deemed necessary to disseminate throughout the empire and remains important today. Papermaking was a pivotal reason Islam spread so quickly. [viii]

754 CE
Abu Ja'far Al-Mansur
City of Baghdad
City of Peace

Abu Ja'far Al-Mansur, the 2nd Caliph (754-775) is sometimes recognized as the real founder of the Abbasid dynasty because he migrated his capital to his new city Baghdad, which he called the 'City of Peace.' With the help of his brother, Abu al-Abbas as-Saffah, he aligned with the Khorasanian rebels and overthrew the Umayyads. When his brother as-Saffah died, al-Mansur became the Caliph. He was a ruthless leader and was one of the main reasons the Abbasid Empire was so powerful. Every Abbasid Caliph descended from al-Mansur. He died while on a pilgrimage to Mecca (Makkah al-Mukarramah) and is buried outside of the city. [ix]

In 762 CE, the Abbasid dynasty, under Caliph Abu Ja'far al Mansur, moved their capital to the newly established city of Baghdad, located on the west bank of the Tigris River in present-day Iraq. Over the next five-hundred years, Baghdad would be known as the world's capital of education and culture, prospering for the entirety of the Golden Age of Islam. The most famous Muslim figures from that period received their educational start in Baghdad or at some time in their careers made Baghdad part of their studies. Of major importance, the city was home to the House of Wisdom, an educational institution that attracted scholars from all over the world. Baghdad was the most affluent and the most scholarly city of the time and was the second largest city in size. Baghdad was the capital of the Golden Age of Islam.[x]

766 CE

Habash Al-Hasib Al-Marwazi

First Calculation of Time

Habash Al-Hasib Al-Marwazi, a renowned Persian astronomer, more commonly known as Habash Al-Hasib, has no absolute records of his birth or death but estimates exist based on his known life records. It is known, for example that he was active in Baghdad in the era of Al-Ma'mun, and then moved to Samarra. The first actual record of him was not until 829/830 CE by Bin Yunus who delineated an observation by al-Hasib. Al-Hasib is credited by Bin Al-Qiftr for a zij (astronomical handbook). Although he is also credited with a less known zij, he is noted with being the author of the third zij, which is almost fully Ptolemaic. In this zij, Al-Hasib says that after Al-Ma'mun's death, he decided to amend the observational data collected by earlier astronomers. Furthermore, Al-Hasib conducted personal experiments about the sun and moon, which ended up creating the first calculation of time, as well as replicated observations of planets and times.

The only remaining zij by Al-Hasib is *The Ptolemaic zij of Habash*, and it is given four names. This zij was in fact used later, such as how Al-Birni used it for is personal astronomical practice. Habash did use other methods, but he had his own methods, some of which came from Indian origin or innovation. Al-Hasib also authored a book titled *Book of Bodies and Distances* which is on five topics of science, including a record of the geodetic speed to ascertain the complete radius of the Earth. Al-Hasib had many works and articles on astronomical instruments. The zij also contains additional trigonometry tables that are beneficial to "the history of trigonometry." Al-Hasib inaugurated "the notion of 'shadow,' umbra (versa)" which is parallel to the modern tangent in trigonometry. It can be concluded that Al-Hasib had a major impact on modern day trigonometry and on our knowledge of time and circumference of the Earth and Moon. [xi] / [xii]

Examples of measurements made by Al-Hasib (still used today)

Earth

- Circumference 32,444 km (20,150 miles)
- Diameter 10,323.201 km (5,414.54 miles)
- Radius 5,161.609 km (3,207.275 miles)

770 CE
The House of Wisdom

During the time of the Umayyad Dynasty, prior to the Abbasid Empire, under Caliph Muawiya 1, a great collection of writings by the greatest writers, polymaths, scientists, theologians, philosophers, and other great thinkers of the ancient world was established in their capital Damascus. The collection was turned into a public library. The library housed Arabic translations of major works originally in Greek, Italian and other scholarly languages. Once the Umayyad Dynasty fell, the Abbasids moved the library to their new capital in the newly established City of Baghdad. Under Caliph Harun Al-Rashid, the library, known as Bayt al-Hikmah, was transformed into the House of Wisdom. After the death of al-Rashid, his son al-Ma'mun continued to support and grow the library. Scholars believe that al-Rashid was responsible for collecting many of the original manuscripts, books and objects from his father and grandfather who had the greatest fortune to rescue them from the ancient libraries of Egypt and Rome; and his son was credited with the translations of the manuscripts into Arabic. Although there may be truth in this, it is also important to know that it originally started during the Umayyad period in Damascus.

Al-Ma'mun turned The House of Wisdom into an extraordinary academy, which became one of the most prominent centers for wisdom during the period and significantly contributed to the scientific revolution occurring during the Golden Age of Islam.[xiii]

During this period, Baghdad was flourishing with diverse cultures, studies, and ideas. The Abbasid Caliphs, who were directly related to the House of Wisdom, included al-Rashid (5th Caliph 786-809), al-Ma'mun (7th Caliph 813-833), al-Mu'tadhid (16th Caliph 892-902) and al-Muktafi (17th Caliph 902-908). Baghdad was rich in wealth and knowledge to the point that leading scholars from all around the middle East would make their journey to study at the House of Wisdom. The leaders of the Abbasid Caliphs and scholars backed the House of Wisdom knew they would find benefactors to support their work. Scholars who arrived at the House of Wisdom varied from authors to scientists. They were not only of Arab cultures but from multiple ethnicities and religions which added to making the culture of the House of Wisdom so rich with different points of view and perspectives on ideas, theories, and knowledge. The House of Wisdom established new traditions in learning; debating on every known topic, to finding out new medical discoveries all of which influenced knowledge from the Golden Age of Islam to the 21st Century.[xiv]

More on the House of Wisdom will be revealed through the many scholars who participated in research, many of which are in this collection.

Nikos Niotis

776 CE

Al-Jahiz

Natural Selection

Abū ʿUthman ʿAmr ibn Baḥr al-Kinānī al-Baṣrī, was born in Basra (present day Iraq) in 776 CE. Much of his life was recorded in a book about authors, *al-Fihrist* (The Catalogue) compiled by Ibn Al-Nadim. Al-Nadim listed close to 140 titles by Al-Jahiz and much of the information provided here can be credited to him.

Al-Jahiz was born into a poor family and worked selling fish in a nearby canal to help his family. Having gone to a madrasa in Basra to start his education he continued his studies both at the madrasa and at his mosque, another source of learning He was a prose writer with works in many areas including literature, theology, zoology, and politico-religious polemics. He reported a membership in the Arabian tribe Banu Kinanah. [2][3][4][5]

Of his many books and focuses of study he was enamored with the study of animals. Close to 1,000 years before the birth of Charles Darwin, Al-Jahiz made his own conclusion that there is an obvious evolution of animals which he explained as their struggle for existence and their transformation of species into each other. This was all explained in *Kitāb al-Ḥayawān* (The Book of Animals). [8] [9]

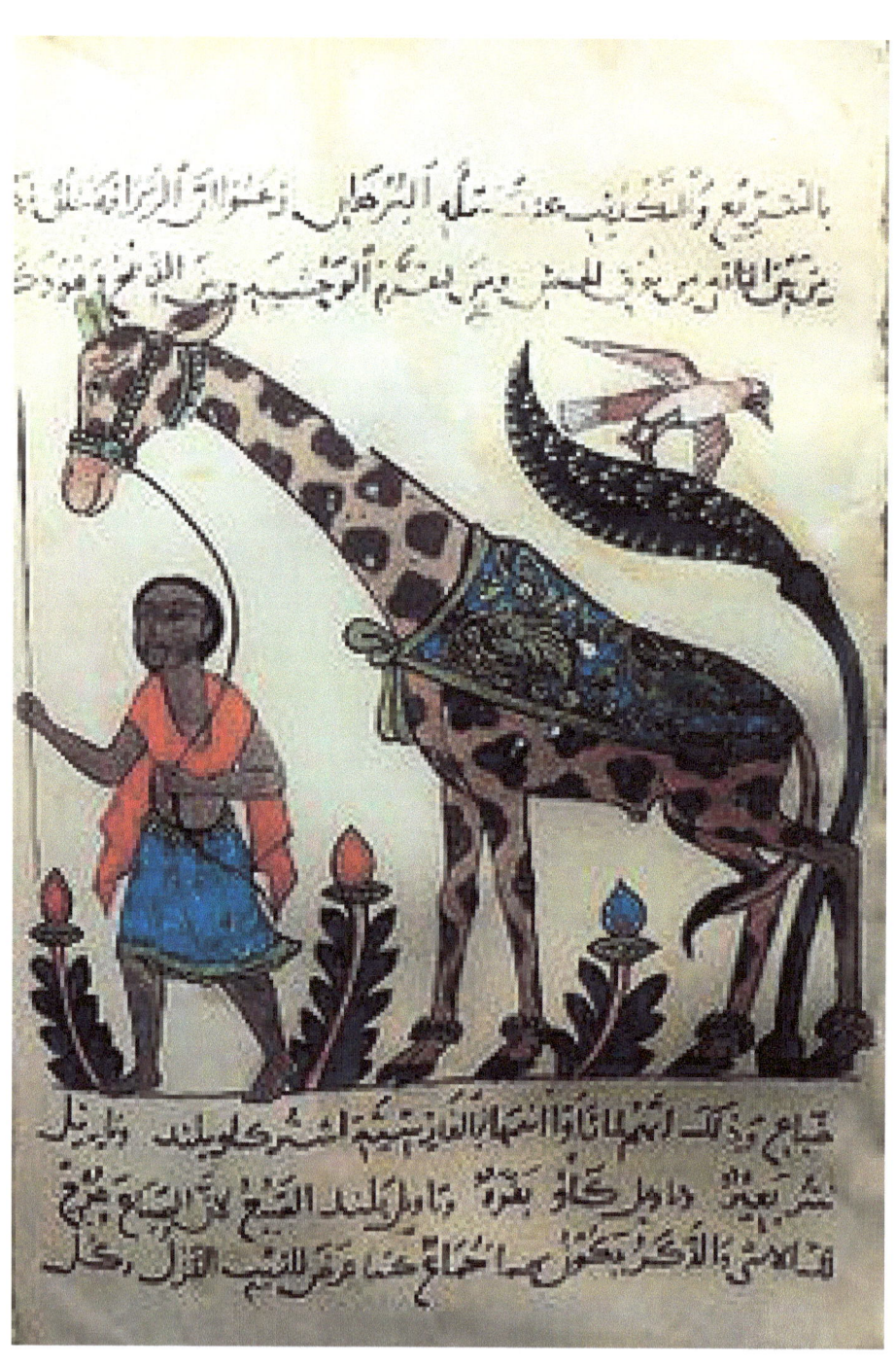

During his early years he was fortunate to have access to translated works from the Greek's including the writings of Aristotle.[19] In 816 he moved to Baghdad at the invitation of the House of Wisdom to join other translators. It was during his time at the House of Wisdom that he wrote most of his books. Under the patronage of the Caliph Al-Ma'mun. The caliph at one time had requested that Al-Jahiz instruct his children but then changed his mind when his children said they were frightened of his boggle eyes. This was later said to be the origin of his nickname, "The Bug Eyed."

His book, *Kitāb al-Ḥayawān* (Book of the Animals) was written as an encyclopedia in seven volumes. It describes more than 350 species of animals. The book was written in honor of his benefactor Muḥammad ibn 'Abd al-Mālik al-Zayyāt, who paid him 5,000 gold coins, likely equivalent to the weight of the collection.[41]

In his book *Kitāb al-Bukhalā'* (The Book of Misers) he authored stories about the greedy. It was written in a humorous and satirical style and is a living example of his prose style of writing. Many of his stories in this collection are published in magazines throughout the modern Arabic-speaking world.[43] [44]

Of special modern-day significance was his book Fakhr Al-Sūdān Ala Al-Bīḍān (Pride of Blacks Over Whites), a debate between imaginary black and white people over which race is superior. Al-Jahiz himself considered himself black. He notes in his conclusion that blacks "have an oratory and eloquence of their own culture and language."[49]

> "Everybody agrees that there is no people on earth in whom generosity is as universally well developed as the Zanj. These people have a natural talent for dancing to the rhythm of the tambourine, without needing to learn it. There are no better singers anywhere in the world, no people more polished and eloquent, and no people less given to insulting language. No other nation can surpass them in bodily strength and physical toughness. ("Internet History Sourcebooks Project") One of them will lift huge blocks and carry heavy loads that would be beyond the strength of most Bedouins or members of other races. They are courageous, energetic, and generous, which are the virtues of nobility, and good-tempered and with little propensity to evil. They are always cheerful, smiling, and devoid of malice, which is a sign of noble character. ("Internet History Sourcebooks Project")
>
> "The Zanj say that God did not make them black to disfigure them; rather it is their environment that made them so." ("Al-Jahiz - Wikipedia") The best evidence of this is that there are black tribes among the Arabs, such as the Banu Sulaim bin Mansur, and that all the peoples settled in the Harra, besides the Banu Sulaim are black. These tribes take slaves from among the Ashban to mind their flocks and for irrigation work, manual labor, and domestic service, and their wives from among the Byzantines; and yet it takes less than three generations for the Harra to give them all the complexion of the Banu Sulaim. This Harra is such that the gazelles, ostriches, insects, wolves, foxes, sheep, asses, horses, and birds that live there are all black. White and black are the results of environment, the natural properties of water and soil, distance from the sun, and intensity of heat. There is no question of metamorphosis, or of punishment, disfigurement or favor meted out by Allah. Besides, the land of the Banu Sulaim has much in common with the land of the Turks, where the camels, beasts of burden, and everything belonging to these people is similar in appearance: everything of theirs has a Turkish look." ("Internet History Sourcebooks Project")[50]

Al-Jahiz returned to Basra after more than fifty years in Baghdad. He suffered from paralysis of the body which may have contributed to his eventual death in January 869. While in his private library, many piles of his own works fell on top of him, killing him instantly.[9]

780 CE

Muhammad Bin-Musa al-Khwarizmi

Algebra and Computer Science

Al-Khwarizmi was a Persian scholar. He was an astronomer, mathematician, geographer, and astrologer. His most notable work introduced algebra into European math. He lived in Baghdad and worked at the House of Wisdom where in 820 he was appointed the head astronomer and director of the library. Of the many accomplishments during his lifetime, he also introduced Hindu-Arabic numbers to the West. He made a set of astronomical tables, based on Greek and Hindu work.[xv] Although al-Khwarizmi was born in Persia, he emigrated to Baghdad, Iraq. Of his many lasting inventions he is most noted as the Father of Algebra as the creator of an algorithm in mathematics which remains in use today.

Even though computer science was not imaginable in the 8th century, he is also called the Grandfather of Computer Science due to his discoveries and inventions that would later influence the foundation of computer science.[xvi] His algebraic contributions is referred to as the basis of sciences. His well-known book, *Hisab al-Jabr wa al-Muqabala* included numerous straightforward quadratic equations.

Arabic Numbers

Al-Khwarizmi also introduced the West to Arabic numbers. While at the House of Wisdom he corrected and organized Ptolemy's research in geography using his own findings in the Qur'an. Near the end of his life, he made tables for sundials so that the time needed to make calculations could be shortened. He died in 850 CE, but not before he invented the shadow square that is now used to determine the linear height of an object. [xvii]

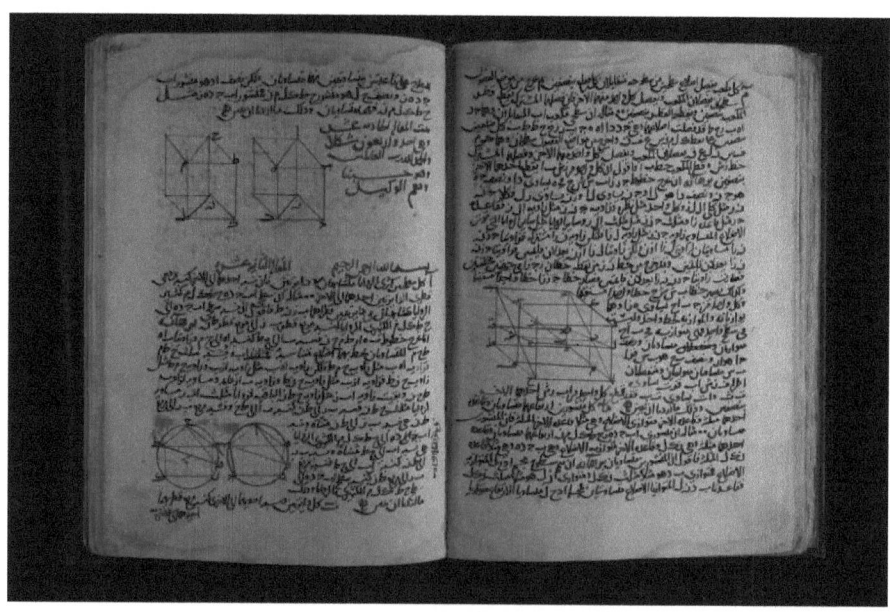

The Father of Algebra

Al-Khwarizmi is credited with the creation of algebra and is called "The Father of Algebra." While a scholar in the House of Wisdom, around the year 825, al-Khwarizmi authored a book called *Hisab Al-Jabr w'Al-Muqabala* (The Compendious Book on Calculation by Completion and Balancing). He solved linear and quadratic equations by balancing either side of the equation where he removed negatives. Although less convenient, he used words rather than numbers or letters as variables. His discovery became a significant contribution to mathematics and most importantly Algebra. [xviii] [xix]

786 CE

Al Ma'mun

Unification of Sunnite and Shi'ite

Al Ma'mun was born in 786 CE, the son of the caliph, Harun al Rashid. Once al Ma'mun came into rule, he resided in Merv and became the 7th Caliph. His life's goal was to end the division of the Islamic World, between the Sunnite and the Shi'ite, something many Islamic leaders have attempted to do for the last 1500 years, unsuccessfully. He was so determined to accomplish his life's goal that he gave his daughter away in marriage to align the two communities. To great dismay, the people of Merv were furious and planned to overthrow al-Ma'mun. When he received word of this, he fled with his powerful vizier to Baghdad. Unfortunately, the vizier was assassinated before they reached Baghdad. When al-Ma'mun reached Baghdad, he decided to end his attempt to unite the Sunnite and the Shi'ite. Over the next fifteen years, al-Ma'mun carefully, but proactively, managed his ministers, as well as not appointing another powerful vizier because of the incident with the last one.

House of Wisdom

Al Ma'mun was a great believer in Greek and Hellenistic philosophy and promoted and encouraged the translation of Greek scientific and philosophical work. Al Ma'mun made his decision to official establish the House of Wisdom as a place to store amazing works of art from around the globe. He also imported manuscripts of important works that did not exist in Islamic countries at the time.

Muslim Doctrine

During his reign, al-Ma'mun developed a liking for science, and established observatories for Muslim scholars to aid them in affirming their astronomical knowledge. Al-Ma'mun decided to make a Muslim doctrine to try and keep things in order. However, the result was the complete opposite. His 'Muslim Doctrine' included the belief in free will, concept of divinity, and human responsibility. The doctrine had an effect for a brief period but ended up crumbling. People from both the Shi'ite and Sunnite did not approve of the full doctrine which sparked anger. While attempting to reverse his doctrine, al-Ma'mun died in August of 833 AD. Scholars say that we "can learn from him to not place our personal thoughts and beliefs into political ideas." [xx]

800 CE
Fatima Al-Fihri
Al-Qarawiyyin Masjid and University

Fatima Al-Fihri was born in 800 CE in Kairouan, Tunisia. She died in 880 CE in Fes, Morocco. Her father, Mohammad bin Abdullah Al-Fihri, was a highly successful businessman. After the deaths of Fatima's husband, father, and brother in short succession, Fatima and her sister, Mariam, received an inheritance of money. They had already received a good education, so they decided to use their money to benefit the community and Fatima used her share of the money to fund the establishment of the al-Qarawiyyin Mosque and University, considered by UNESCO and the Guinness Book of World Records, as the oldest continually operating degree granting university in the world.[xxi]

Fascinated by the art and philosophy behind learning new things and inventions, she insured that the university taught various religious and physical sciences, law, medicine, Arabic grammar, and calligraphy. Fatimah al Fihri's University started a whole new era in formal teaching.[xxii][xxiii] / [xxiv] / [xxv]

801 CE

Al-Kindi

First Muslim Philosopher

Abu-Yusuf Ya'qub Bin Ishaq Bin as-Sabbah Bin 'Omran Bin Isma'il al-Kindi, was born in 801 CE. He was born in the city of Kufa where he received the finest education available, after which time he moved to Baghdad, where he continued his studies. Once in Baghdad, al-Kindi became well-known for his knowledge and caught the attention of Caliph al-Ma'mun. After al-Ma'mun's passing, Al Kindi continued at the House of Wisdom until his philosophical views became competitive to the views of other influential scholars forcing his departure.

Al-Kindi was an Arab Muslim philosopher, polymath, mathematician, physician and musician, pharmacist, ophthalmologist, physicist, geographer, astronomer, and chemist. As the first philosopher with pure Arabian blood, he is considered the first Muslim philosopher. The renowned philosopher wrote on many scholarly subjects including mathematics:

- Indian numbers
- Linear multiplication
- the "harmony of numbers"
- measuring proportion (space) and time
- relative quantities
- numerical procedures
- and cancellation.

Al-Kindi authored many works of philosophy such as *That There Are Incorporeal Substances* and *Discourse on The Soul*. These works dove into the immaterial nature of the soul of a human. Al-Kindi applied math to other fields, such as optics, by drawing geometrical diagrams to describe light rays and the notion of refractions, shadows, and reflections. His ideas also impacted the renowned mathematician Bin Al-Haytham. Al-Kindi was employed to translate Greek texts at the House of Wisdom. He described and wrote about how the theory of parallels, such as how Aristotle wrote of a mirror that was used to set a ship on fire during a battle. [xxvi] / [xxvii]

Frequency Analysis

Al-Kindi authored a treatise, *A Manuscript on Deciphering Cryptographic Messages* including a description on his method of frequency analysis. Al-Kindi died in 873 CE with over 260 published books including books on the topics of space, math, and philosophy.[xxviii] / [xxix]

803 C.E.

The Banu Musa Brothers

"Sons of Moses"

The Banū Mūsā Brothers ("Sons of Moses"), Abū Jaʿfar, Muḥammad Bin Mūsā Bin Shākir (d. February 873), Abū al-Qāsim, Aḥmad Bin Mūsā Bin Shākir (d. 9th century) and Al-Ḥasan Bin Mūsā Bin Shākir (d. 9th century), were three 9th-century Persian[3][4] scholars who lived and worked in Baghdad. Their father, Mūsā Bin Shākir, who earlier in life had been a highwayman and astronomer in Khorasan died when they were young, but Caliph Al-Ma'mun recognized the Banu Musa brothers for their talents and mental ability. Subsequently, the brothers were enrolled in the House of Wisdom, and they excelled in mechanics, mathematics, and astronomy.

They are best known for their *Book of Ingenious Devices on automatic machines and mechanical devices*. Also significant was their treatise, The *Book on the Measurement of Plane and Spherical Figures*, a foundational work on geometry that was frequently quoted by both Islamic and European mathematicians.xxx

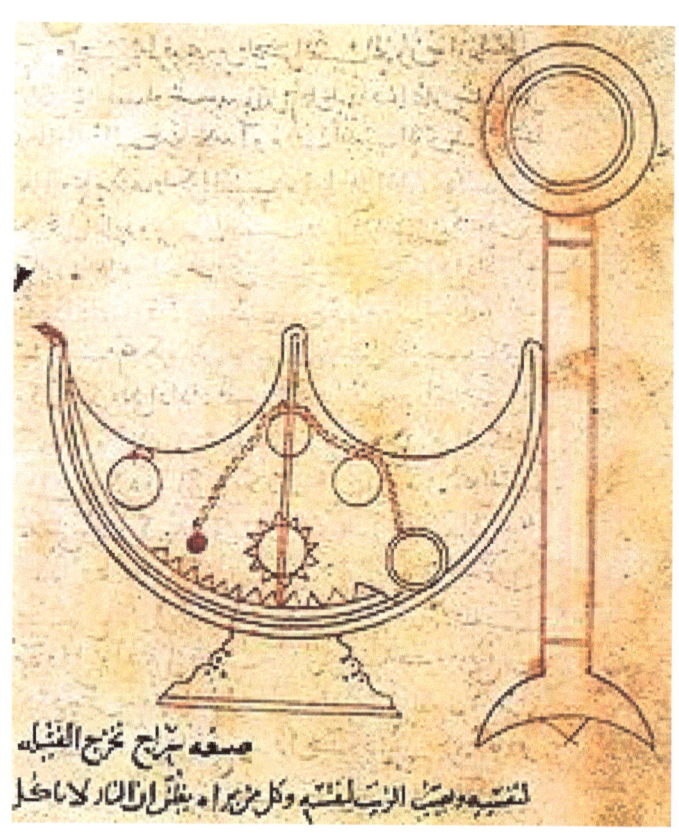

While at the House of Wisdom, under the leadership of Yahya Bin Abi Mansur,[6] they led efforts in translating ancient Greek works into Arabic by sending for Greek texts from the Byzantines, paying large sums for their translation, and learning Greek themselves.[7] One story on how translations were paid for, translators were paid in an equal amount of gold based on the weight of the work translated. On one such trips, the Ban Musa brother Muhammad met and recruited the famous mathematician and translator Thabit Bin Qurra. At some point Hunayn Bin Ishaq also became part of their team. The Banu Musa brothers sponsored many scientists and translators to work at the House of Wisdom with month sponsorships of 500 dinars. It was because of their extensive work translating ancient texts that they exist today. Another extremely important fact, not to forget, is that Europe fell into the Dark Ages at the same time as the Golden Age of Islam. Modern scholars can thank the scholars of the House of Wisdom for saving texts that would have otherwise been lost when the libraries were burned down by the barbarians who attacked Europe after the Renaissance.

As the brothers translated works, they observed what they considered to be incomplete arguments by the Greek philosophers. An example of this was in their treatise, *Kitab Marifat Masahat Al-Ashkal* (The Book of the Measurement of Plane and Spherical Figures) where they added volume and area number values leading them to establishing their own original works. Another popular work was the *Kitab al-Hiyal* (The Tricks Book) which is credited to the middle brother Ahmad. This book included more than 100 inventions many of which are still popular in the 21st Century, more than a millennium since their discovery. Three of these inventions include a lamp that would mechanically dim, alternating fountains and a clamshell grab, the 3rd of which is owned today by most dog owners. Some of their inventions were based on creations by earlier Greek inventors but embellished and modernized.

Automatic Musical Instruments

The Banu Musa Brothers are credited with inventing a hydro powered organ that used exchangeable cylinders with pins; an automatic flute playing machine that used steam power, and the first programmable music sequencers.[44] [45] [46] [44] [47]

Throttling Valve

Included in their *Book of Ingenious Devices* was a throttling valve with variable structure control. This device offered two-step level controllers for fluids, a form of discontinuous variable structure controls.[62] / [63]

Other Inventions

- Automatic controls[11]
- Automatic crank.[16]
- Valves [13][14] [13] [17]
- Automatic fountains .[3]
- Mechanical musical machines.[22]
- Practical tools including the mechanical grab (clamshell grab) [14]
- Water dispensers for hot and cold water.[3]
- Mechanical trick devices[14]
- Hurricane lamp[14]
- Self-trimming lamp[14]
- Self-feeding lamp[14]

809 CE
Hunayn Bin Ishaq
Ophthalmology

Hunayn Bin Ishaq was born in Al-Hirah (ancient city in Mesopotamia located near modern day Kufa, Iraq) in 809 CE and died in Baghdad in 873 CE. He was a Christian scholar, physician, scientist, and translator born to an Arab family. He was also known as Jonnatius. Hunayn had a son, Ishaq and a niece named Hushayn.

He had a close relationship with al-Mutawakkil (10th Caliph 847 – 861) and the Caliph made him his personal physician. Hunayn studied in the field of Ophthalmology which is the study of eye surgery and eye disorders. He wrote his first book in this field, which is called *Ten Treatises on Ophthalmology*. Hunayn also wrote *How to Grasp Religion* from another area he studied. Jonnatius was fluent in Arabic, Greek, Syriac, and Persian. He was mostly known as a translator of medical books which were used by medical students and doctors. He also translated some of the philosopher's works of Aristotle, and Plato which he translated from Greek to Syriac. His children further translated his works from Syriac to Arabic and Hunanyn would help correct and fix mistakes and errors.[xxxi]

Hunayn became the leading translator in the House of Wisdom, and he was especially famous for his Arabic translations. Hunayn was exceptionally good at languages when he was younger, learning Arabic and Syria. During his lifetime he traveled to Alexandria where he excelled in the Greek language, translating Greek texts into Arabic which spread to Mesopotamia, Syria, and Egypt. [xxxii]

825 CE
Taqi al Din
Al-Shammisiyyah Observatory

The Al-Shammisiyyah Observatory, built in Baghdad in 825 CE, was the first observatory that specialized as a research institute.[50][51][52]

حاج شیخ مرتضی انصاری

835 CE

Ahmad Bin Tulun

Economist

Ahmad Bin Tulun wa born in September of 835 CE and as a child, Ahmad was taken into slavery and placed into the private service of the Abbasid caliph in Baghdad. When he got older, he eventually gained power in the Egyptian government, where he would observe how the minister of finance struggled in controlling his department. He eventually became vice-governor and gained many successes for the country. With this power he took control of the Ministry of Finance and was able to improve the financial state of Egypt with his economic policies. He was able to compound tax revenues by increasing the agricultural output, and the outcome of this financial plan was very successful. He also built a mosque in 879 named The Mosque of Ahmad Bin Tulun in Cairo, Egypt, and is the oldest standing mosque in the city today. He died in 884 CE.[xxxiii]

His life's story closely resembles the story of Joseph in the Old Testament.

872 CE
Al-Fustat Hospital
Bimaristan

Bin Tulun established the first known mental institute in Fustat (modern day Cairo) in 872 CE. The hospital operated for approximately 600 years. It was like other hospitals in that it had separate bathing areas according to gender. A free hospital, it was the first known hospital to take care of mentally ill patients. It housed a teaching hospital, an emergency department, a pharmacy, and a large medical library housing upwards of 100,000 volumes. A little known fact about the mental health treatment was their using music therapy to treat the insane.[39] [31]

879 CE

The Mosque of Ahmad Bin Tulun

The Mosque was built in the year 879 in Cairo. It is rumored that the location is where Noah's Ark came to rest. In addition to this claim, local lore says that a large Qur'anic verse found in the mosque is written on wood from the Ark. The design of the mosque was greatly influenced by the courtyard of the Prophet where he spoke to his followers in Medina. The prayer halls are all designed to give visitors the feeling of being in their own living rooms. In the center of the mosque is a sphere showing direction to Mecca for worshippers in the mosque.

836 CE

Thabit Bin Qurra
Trepidation of the Equinoxes

Thabit Bin Qurra was born in 836 CE in what is now Turkey. Thabit was part of the Sabian religious sect and had traditions of Hellenistic culture, worshipped nothing except the stars, and had a very strong knowledge of Greek and Arabic. Because of this knowledge he was invited to the House of Wisdom by Muhammad Bin Musa Bin Shakir. After traveling to Baghdad when accepting the invitation, Thabit quickly rose to fame in mathematics, medicine, philosophy, astrology, and mechanics. In fact, the early astronomical theory of the trepidation of the equinoxes comes from Thabit. Furthermore, Thabit was able to ascertain the length of a year as 365 days, as stated by Copernicus. Thabit also rejected Aristotle's ideas on the immobile. Thabit is also somewhat seen as the originator of statics in the fields of mechanics and physics. He went on to author papers about Archimedean problems in statics and mechanics . Although not all Thabit's books survived to this day, the majority are on the topic of math. In his books, Thabit placed the basis to continue to work on the examination and regeneration of Ptolemaic astronomy. Thabit Bin Qurra died in 901 CE. [xxxiv]

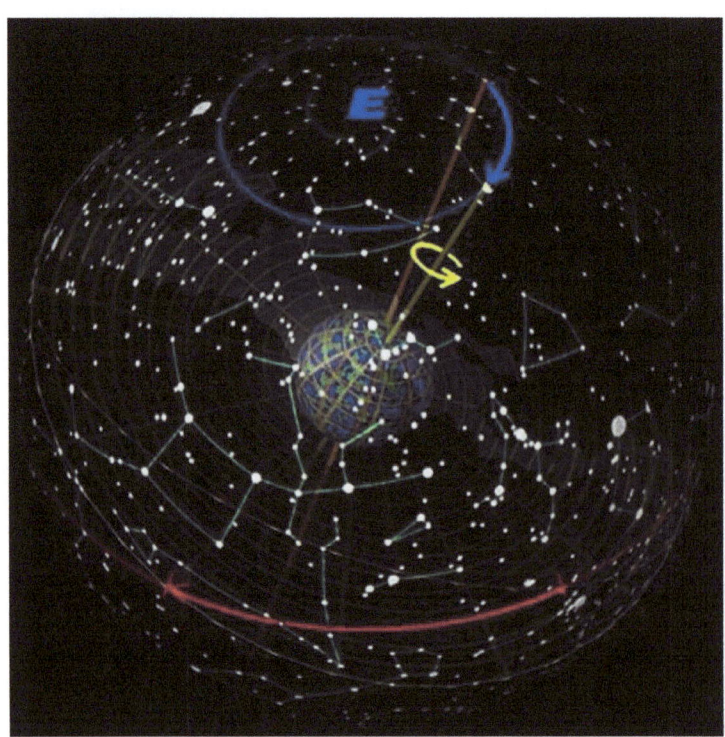

840 CE

Al-Adli ar-Rumi

Chess Manual

The oldest known chess manual was in Arabic and dates to 840–850, written by Al-Adli ar-Rumi (800–870), a renowned Arab chess player, titled Kitab ash-shatranj (*Book of Chess*). During the Islamic Golden Age, many works on shatranj were written, recording for the first time the analysis of opening moves, game problems, the knight's tour, and many more subjects common in modern chess books.[27]

858 CE

Al Battani

Astronomical Calculations

Al Battani was born in Harran (Carrhae, Turkey) in 858 CE, then moved to Raqqa, Syria where he received his education to become a scholar. He then moved to Samarra to work. Battani had been introduced to the stars from an early age. When his family lived in Harran, his family were star worshippers, astronomers. Al Battani became a Muslim later, but still held on to astronomy. Battani was a huge help in cataloging the stars because he cataloged 489 of them. He also calculated the exact time in a full year which is 365 days, 5 hours, 46 minutes, and 24 seconds. Today the time of a year is rounded to 365 days, and a quarter.

Al Battani also did the same for all the seasons. In the equinoxes, he is the one who calculated 54.5" per year. He also calculated the inclination of the ecliptic, which is 23 degrees, and 35'. Al Battani did not only influence Astronomy, but also math. In his calculations he used trigonometric methodology, when others use geometry for calculations in science. He died in 929 C.E. in Iraq leaving behind a legacy that will be remembered. [xxxv]

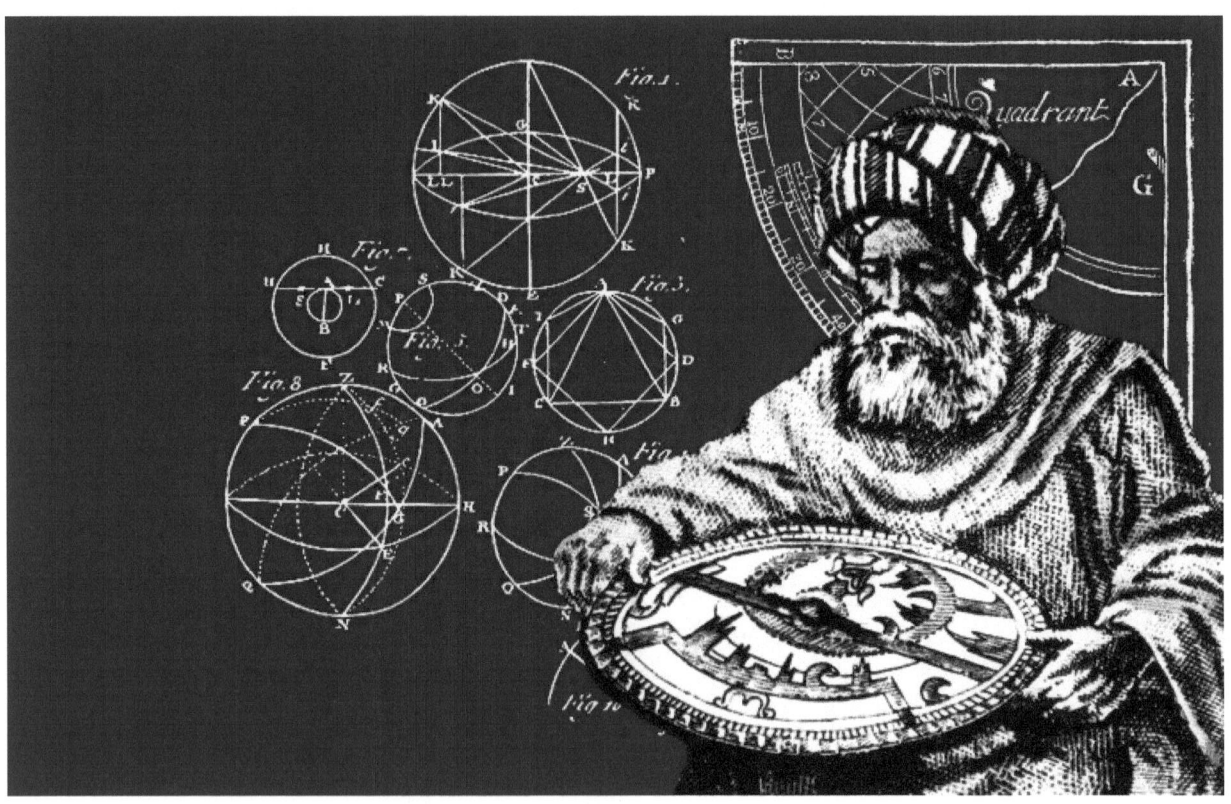

864 CE
Abū Bakr Muhammad Bin Zakariyyā al-Rāzī
Differentiated Measles from Smallpox

Abu Bakr was a Persian polymath, physician, alchemist, philosopher, and important figure in the history of medicine. He also wrote on logic, astronomy, and grammar. He was known to have been one of the greatest physicians of the Islamic world. He was born 864 CE in Ray, Iran and died October 15, 925 CE also in Iran. Of the countless discoveries he made in his lifetime, he described measles and smallpox as separate diseases and wrote the first book on pediatrics. He also developed a metaphysical system, which proposes that the universe consists of five elements: God, time, place, soul, and matter.[xxxvi]

Medicine, Kerosene and Soap

Abu Bakr worked at the leading hospitals in Rayy and Baghdad and at the Samanid court in Central Asia. His most renowned book is titled *The Comprehensive Book on Medicine*. This work includes previous authors thoughts on diseases, as well as his own. His other main work *Book of Medicine Dedicated to Mansur* focuses on medicine and was referred to for more than 600 years after this death.[xxxvii]

Kerosene

Abu Bakr invented a flammable liquid, an oil that can produce light. In 1854 a man named Abraham Genser named it Kerosene. [xxxviii] Although the Chinese used Kerosene by extracting and purifying petroleum, the process of distilling crude oil into kerosene, it was first written about in Abu Bakr's *Kitab al-Asrar* (*Book of Secrets*).[40][41] He also referred to a Kerosene Lamp in his book which he called the "naffatah". [42]

Soap

In the 9th century, Abu Bakr developed hard toilet soap to wash away impurities from their hands and bodies. This substance is still used for bathing and housekeeping. Although there are records of soaps as early as 2200 BC in Babylon, Abu Bakr discovered that water did not completely take away the microbes and bacteria and that soaps are basically the salts of fatty acids.[xxxix] Abu Bakr also had a recipe for producing glycerine from olive oil..[38]

"I have written 20,000 pages; moreover, I spent fifteen years of my life – night and day – writing the enormous collection entitled Al Hawi. It was during this time that I lost my eyesight, and my hand became paralyzed, with the result that I am now deprived of reading and writing. Nonetheless, I have never given up."

~**Abu Bakr Muhammed** [xl]

870 CE
Abu Nasir al Farabi
Logic, Physics, Music, Metaphysics, and Politics

Al Farabi was born in Baghdad in 870 CE, but not much was recorded about his life until almost three centuries after his death. He was known historically for his life of teaching, studying, and writing. Many of Al Farabi's works were on the topics of logic, physics, music, metaphysics, and politics. His books *The Political Regime* and *Virtuous City* discussed politics and differentiated between governments.

In his other work, *The Selected Aphorisms,* Al Farabi described how the ancient people would teach, such as Aristotle, Socrates, and Plato. This work included how Islam viewed things such as the afterlife, the structure of the world, war, and succession. Another work titled *Enumeration of the Sciences* primarily focuses on metaphysics (a branch of philosophy) and logic. Additional works by Al Farabi includes *Book of Letters* that probes on language, translation of work, and logic. Finally, Al Farabi's book *Kitab ihsa al-ulum* centers on how much Aristotle stresses on knowledge, displaying Al Farabi's own beliefs. Al Farabi died in 950 CE, in Damascus, Syria. [xli] / [xlii]

875 CE

Abbas Ibn Firnas

The First Human Flying Machine

In the year 875 Abbas Ibn Firnas became the first man to fly, although landing was not an accomplishment. Ibn Firnas was a physician, engineer, and inventor. He was first inspired by Armen Firmen who is credited with creating the first parachute. When Firnas saw him demonstrate his parachute in a crowd, he was influenced to try something more exciting, to fly. Not to be confused with the Greek Myth of Daedalus and his son Icarus, Firnas created a pair of wings out of silk, wood, and feathers. He climbed to the top of the minaret of the Grand Mosque of Cordoba and after jumping managed not to kill himself when he crash landed. The original apparatus slowed his decent somewhat like a parachute.

Although he failed to achieve his repeated attempts to fly, he continued trying for more than a decade. Twelve long years after he started his experiments, many of which resulted in personal back injuries, that although tolerable would haunt him the rest of his life, he realized that he was missing something that could help him land slower. Seeing how a bird used their tail to glide downward slowly, he created another flying machine that worked, at least for a 10-minute slow decent to the ground. Another important invention credited to Ibn Firnas was the manufacture of glass from stones.[36] Abbas Ibn Firnas died in the year 887 (birth unknown). xliii

1000 years later, The Wright Brothers, in their first flight at Kitty Hawk, NC on December 17th, 1903, only stayed afloat 12 seconds.

879 CE
Ahmad ibn Fadlan
Diplomat, Traveler and Viking Expert

Ahmad ibn Fadlan ibn al-'Abbas ibn Rasid ibn Hammad was known as an expert in jurisprudence and faith and was part of the court of al-Muqtadir (18th Caliph). Little is known about his life until 921 when the Caliph sent him to be the secretary to the Abbasid ambassador located in Volga Bulgaria, Almis. The mission to Volga Bulgaria was led by Susan al-Rassi, a eunuch in Caliph al-Muqtadir's court, who's instructions were to explain Islamic law to a recently converted Bulgar peoples living on the Volga River in what is now Russia.

The diplomatic party established new caravan routes from Baghdad to Bukhara, now part of Uzbekistan. Not having already established direct routes, the delegation covered 4,000 kilometers (2500 miles) and contacted many different peoples including Khazar Khaganate, Oghuz Turks, lands on the east coast of the Caspian Sea, the Pechenegs on the Ural River and most importantly the Varangian (Vikings).

Ibn Fadlan established important trade relations with the Varangians and recorded the first known account of a Viking funeral, in which the Chief of their tribe was cremated. What affected ibn Fadlan most about this ritual was that one of the chief's 3 main slaves volunteered to be cremated with her chief believing she would marry him in Valhalla. This ritual included intoxicating her following by strangulation and stabbing her to death. Once deceased, the slave girl and her chief were placed in his favorite Oseberg (Viking ship) and burned.

882 CE
Saadia Gaon
Jewish Philosopher and Theologian

Saadia Gaon was born in Cairo, Egypt in 882 and died in Baghdad in 942. Otherwise known as Rasag, he was a philosopher, poet, biblical commentator, polemicist, grammarian, and pre-eminent Jewish leader of the Mesopotamian Jewry. He received his education in Cairo. In 928, Saadia became leader of the Academy of Sura, and his influence spread throughout the Jewish world. He was known for his work on Hebrew Linguistics, translations of Hebrew texts into Arabic, Jewish law, and the concept of preventing a schism in Judaism by means of simple argument.[xliv]

Of his many writings he wrote two Arabic translations of the bible and his own version of the Hebrew dictionary, poems, and a Jewish prayer book. .One of Saadia's more famous works is *Kitab al-amanat wa al-itiqadat* (The Book of Beliefs and Opinions.) [xlv / xlvi]

893 CE

Yahya Bin Adi

Theorist and Doctor

Yahya Bin Adi, born in Tikrit, Iraq in 893. He was a theorist in logic and a doctor. Although born in Tikrit he later moved to Bagdad at a very young age to get his formal education, where he learned logic and philosophy alongside other well-known logicians of the day. Using his Arabic and Greek language skills, he translated dozens of Greek philosophies into Arabic. He died in 974. [xlvii]

896 CE

Ali Bin Al Husayn Al Masudi

Herodotus of the Arabs

Al Masudi was born in Baghdad in 896 CE and died in Cairo in 956 CE. Like the Ancient Greek Herodotus, Al-Masudi was a great traveler who recorded history and important persons of his time. Some of his known destinations include Basra, India, Ceylon, Zanzibar, Oman, and Aleppo. Al Masudi authored thirty works. Unfortunately, only two of the works he wrote survived, *Book of Notification and Review* and *Meadows of Gold,* his most famous writing.[xlviii]

903 CE

Al Sufi

Nebulosity of the Nebula

Al Sufi was one of Islam's greatest astronomers. He was born December 7, 903 in Rayy, Iran. His most notable literary work was about constellations, which he finished around 964. His book included and described the 'nebulosity' (cloudiness) of the nebula. He identified stars, galaxies, the Milky Way, ecliptic plane and more, focusing more on the Southern skies. He made many astronomical drawings, one from the outside of a celestial globe, and the other from the side. He also wrote about the astrolabe and its 1000 different uses. He died in 986 CE. [xlix / l]

920 CE
Abu'l-Hasan al-Uqlidisi

Decimal Fractions / Arabic Numerals

Abu'l Hasan Ahmad ibn Ibrahim Al-Uqlidisi, actual dates of birth and death are not known, however historians believe he was likely born near 920 and died in 980. He was a mathematician notable for his calculations without deletions using decimal fractions. He is also the earliest known mathematician and scientist to use Hindu (Arabic) numbers. His earliest surviving treatise is called the *Kitab al-Fusul fi al-Hisab al-Hindi* (The Arithmetic's of India)."

A more modern Persian mathematician, Jamshid al-Kashi, claimed he discovered the use of decimal fractions in the 15th century only to be proven a fraud by historians familiar with al-Uqlidisi.[4] [3] [69][70]

932 CE

Ali Abuzar Mari

The Fountain Pen

In his book, *Kitab al-Majalis wa 'l-musayarat*, the Fatimid caliph Al-Mu'izz li-Din Allah demanded his scientists and inventors to create a pen that would not stain his hands or his clothes. The task was taken on and accomplished by Ali Abuzar Mari.

Mari created a fountain pen where the ink was held in a reservoir and it allowed the pen to be held upside down without leaking.[72]

933 CE
Al-Hakim Nishapuri
Imam of the Muhaddithin

Al Hakim was a poet and a writer. He was born March 3, 933 CE. He died September 1, 1014, CE. He was sometimes called the Imam of the Muhaddithin. When he was 72 years old Al Hakim wrote *Al Mustadrak Alaa Al-Sahihain.* [li] / [lii] / [liii]

935 CE

Abu Qasim Ferdowsi

Persian Poet

Known as Ferdowsi lived from 935 CE to 1026 CE. He was born in Tusi. His most famous book *"Shah nameh"* was written in Persian with a little bit of Arabic. The title of this book means *"Book of Kings."* The book is a poem of sixty thousand couplets. He authored the book to make enough moneys to give his daughter a dowry. It took him 35 years to complete and is considered the longest poem ever written.[liv]

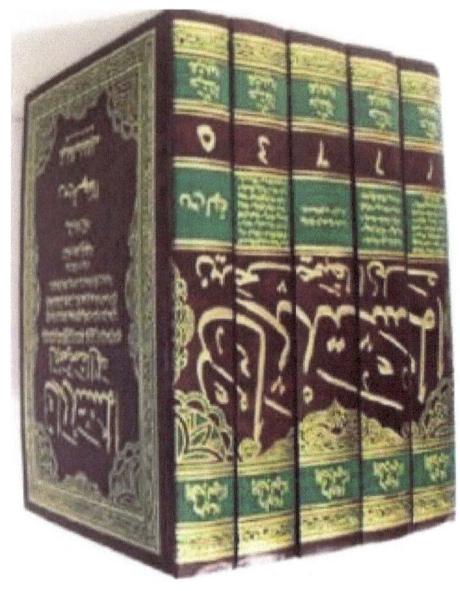

936 CE
Al-Zahrawi
Inhalation Anesthesia

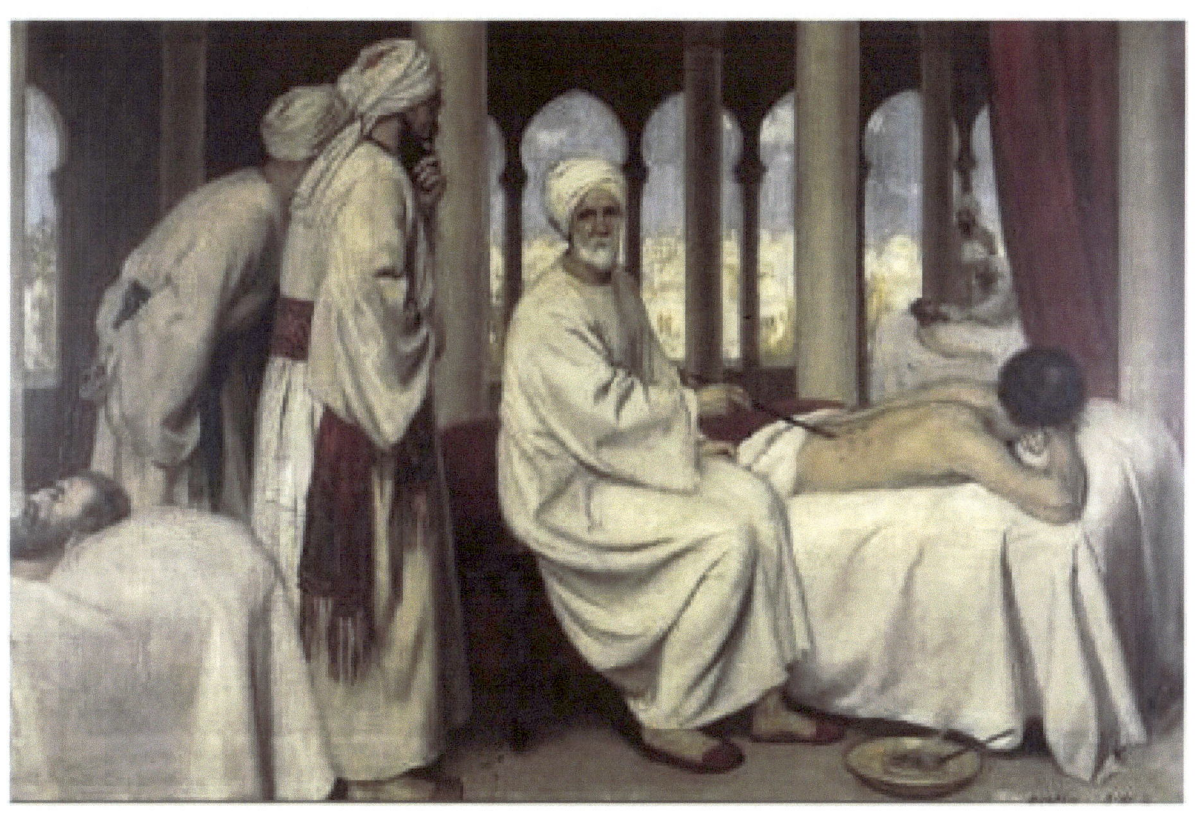

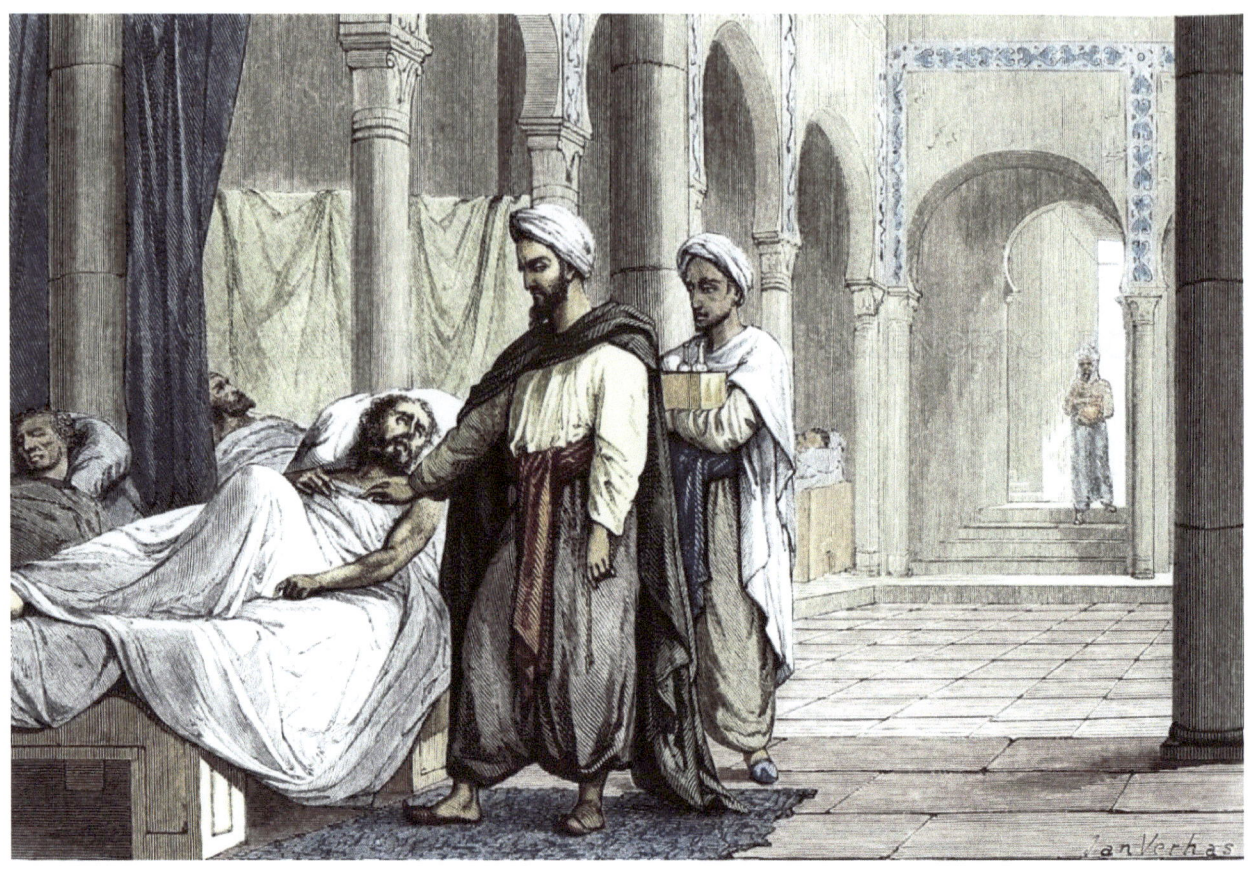

Al-Zahrawi was born in the year 936 CE and died 1013 CE. He was born in southern Spain. At the time of his birth Spain was part of the Islamic empire. Al-Zahrawi was one of the greatest surgeons of his time. Both he and Al-Razi Bin Sina were very famous for their partnership in the medical field. They both created more than 200 surgical instruments that were used for nearly 500 years, and some are still used today although modified to our timeline. Al-Zahrawi wrote *On Surgery and Instruments* a 30-volume illustrated work, which is about 1500 pages in length. It serves as an encyclopedia all about medicine, pioneering, and surgery. His work included many topics like midwifery, pharmacology, therapeutics, dietetics, psychotherapy, weights and measures, and medical chemistry. He was the personal physician of Caliph Al-Hakam-II. He was considered the father of operative surgery. [lv]

Surgical Instruments in the 9th century were not considered capable for intricate types of medicine, which in turn limited it is advancement. At least not until Al-Zahrawi decided to make his own. Al-Zahrawi made these instruments because he was a specialist in the surgery department himself and had firsthand knowledge of the limitations of existing technology. The surgical instruments that they had were bone saws and fine scissors for eye surgery. These instruments were very handy in his day and are still known today, however modernized. Some of the inventions include scalpels, specula, pincers and lithotrites that are used for crushing bladder stones.[lvi]

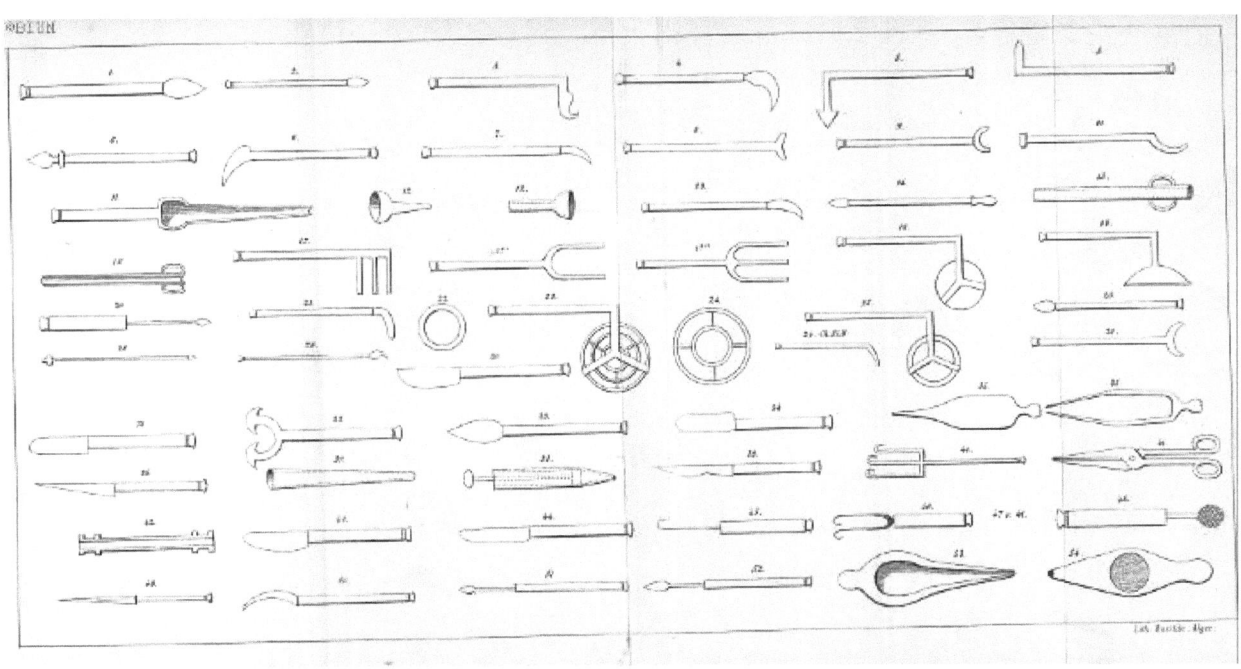

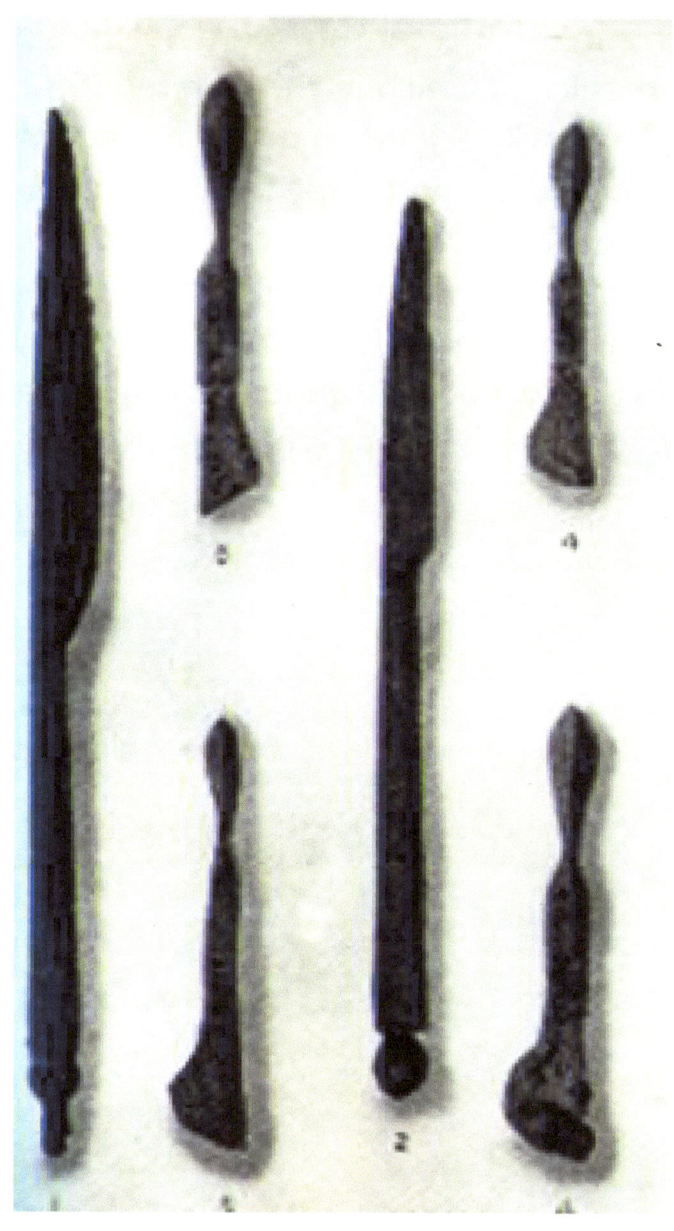

Inhalation Anesthesia

Perhaps one of his greatest inventions was inhalation anesthesia. He used a sponge dipped in a variety of medicines and applied the sponge to the patients nose and mouth, invariably making the patient unconscious.

940 CE
Abu Al Wafa Buzjani
Astronomer

Abu Al Wafa Buzjani was born in 940 in Buzjan, Iran. Buzjani was a major astronomer as well as a very important mathematician. The period of 975 thru 985 were Buzjani's most successful years. When he moved to Baghdad, he started working at an observatory called Bab Al Tobin. Reportedly this observatory was destroyed several years after Buzjani completed his many observations. Buzjani and Al-Biruni worked together in observing the lunar eclipse, which led to differentiating times in different zones. Buzjani also helped Biruni in many of his calculations. Buzjani had a passion for writing, and wrote twenty-two books about astronomy, geometry, and trigonometry. Unfortunately, only eight of these books survived. One of his famous books entitled *Kitab Al Majisti* helped many future scientists establish their own forms of calculation. This book had three chapters on trigonometry, astrology, and planetary theory. Abu Al Wafa Buzjani was a very important individual and made major contributions to both arithmetic and astrology. He died in 997. [lvii]

944 CE
Mariam Al-Astrolabiya
Astronomer

Mariam Al- Ijliya Al- Astrulabi was born in 944 in Syria. She studied an instrument called the astrolabe. It was a very complicated device used to track the sun and stars to tell time and to identify direction. Astrolabes are an ancient astronomical computer for solving problems relating to time and position of the sun and stars. It was Mariam's academic brilliance and an exceptionally focused mind that lay the foundation for the transportation and communication we see in the modern world. Making astrolabes, a branch of applied science of great status, was practiced by many women from Aleppo (Syria).

She added adjustments to the Astrolabe making it more accurate. This device was also used for religious benefits, such as knowing what time it was to pray, or when Ramadan would start and end. Most important, the astrolabe determined which direction to Mecca for prayer times. [lviii]

She is known to be the greatest female astronomer in ancient Islam. She died in 867. [lix]

Astrolabe

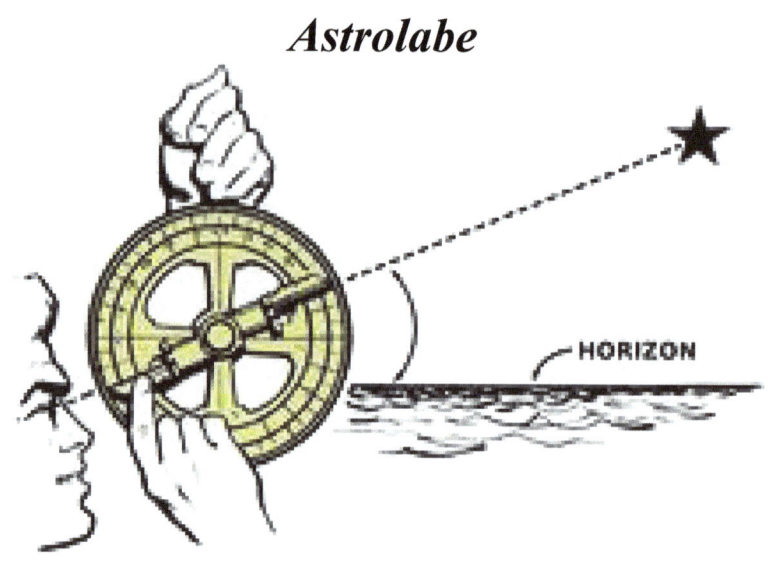

Astronomers used the astrolabe to track the stars. It was invented around 200 CE by an astronomer in Greece. It can also find the positions of the sun, moon, and planets. During the Golden Age of Islam, it was also used as a qibla (compass) and to calculate times to pray. The astrolabe was improved upon by Mariam Al- Ijliya Al- Astrulabi, who was able to calculate when an eclipse would happen. The astrolabe is still valued and used to this day. [lx]

Back in the 9th century the astrolabe was like a smartphone of their generation. [lxi]

946 CE
Muḥammad Bin Aḥmad Al- Muqaddasi
Geographer

Al-Muqaddasi, born in 946 was an Arab geographer. He had an extensive education, and after travelling to Makkah (Mecca), he decided to study geography. Eventually, he authored a book where he split the Islamic world into 14 regions, and he explained each one separately. The book was presented in two main parts: describing the region, and the second part including the population, diversity, social classes, political groups, and much more. He died in 1000. [lxii]

965 CE

Bin Al-Haytham

The Camera Obscura

Al-Haytham (Abu Ali al-Hasan Bin al-Haytham) was a mathematician and a scientist. He was born July 1, 965 CE in Basra (modern day Iraq) and died March 6, 1040. Bin Haytham was first appointed Minister of Basra, a job he became very disillusioned with.

He had great interest in science and, after resigning his position, travelled extensively around the Middle East to research everything he could learn about light and optics. He chose to stop and remain in Cairo, Egypt where he would spend the rest of his life.

Bin Haytham wrote three major works, one of them an autobiography that took at least 100 versions, the second was the Book of Vision, and his third was an investigation of Euclid's elements. One of his largest attempted projects, one that he could never complete, was to slow down the flow of the Nile river. This attempt gained the notice of the Caliph Al-Hakim who made him the head of an engineering team to lead this project on behalf of the Caliph. Bin al Haytham eventually determined he could not complete the job as his hypothesis was wrong. The Caliph made it known that he was disappointed in this result.

With the knowledge that the Caliph had a reputation for murdering those he was not pleased with, feigned a serious illness, and stayed inside of his home until the Caliph himself died. Once the Caliph was no longer a threat, Bin al-Haytham reemerged and restarted his research and experimentation.

The Eye

He experimented on how the eye works. He understood how the human eye can transmit light and create vision. Bin al Haytham conducted an experiment in which he went into a dark room and created a hole in the wall which let light come into the wall in front of it. With the light projecting on to the wall he concluded many well-known facts of the scientific vision method of today. During the years that Bin al Haytham lived in Egypt he invented the Camera Obscura (also called the Pinhole Camera) and authored his Book of Vision after this experiment. This brilliant scientist helped future generations to understand more about light, optics, and vision than ever before. [lxiii]

During his lifetime, Bin Al-Haytham wrote ninety-two pieces in which fifty-five survived. The topics range from astronomy, mathematics, and optics. Bin Al-Haytham authored a seven-volume book on optics, and in Book 1, Bin Al-Haytham states he will use real evidence rather than simply a theory to prove his investigation on light. In terms of mathematics and number theory, Bin Al-Haytham used congruences which is now named Wilson's theorem. In Bin Al Haytham's work *Analysis and Synthesis*, the mathematician is focused on studying methods from numerous mathematicians to solve equations. Although the ancient Greeks did use analysis to solve geographic problems, Bin Al-Haytham used a more general method that can be applied to algebraic problems. The mathematician died in 1040 CE in Cairo.[lxiv][lxv]

The Camera Obscura

The pinhole camera, or camera obscura, was invented by Bin Al-Haytham, the man who figured out how our eyes work. Before around the tenth century, people believed that if one wanted to see an object, they must open their eyes and send out a beam of light, and that light would bounce back, giving them an image if an object. However, Bin Al-Haytham proved that this was incorrect. He said that when one opens one's eyes, light streams in, allowing one to see the world around them. He observed that light coming through a tiny hole travelled in straight lines and projected an image onto the opposite wall. "Based on this experimentation, Bin al-Haytham concluded that vision is accomplished by rays coming from external bright sources and entering the eye, rather than through rays released from the eye was commonly believed." ("Islamic heritage series: Ibn Al-Haytham: The pioneer optician") Scientific knowledge today is a result of thousands of years of research. This information has been collected by people with different and diverse languages religions, and cultures.[lxvi]

Al-Haytham was one of the first to conduct experiments to evaluate his theories. To do this, he built the camera obscura, a dark room with a screen and a small pinhole to let light in. Whatever was outside of the pinhole would be projected on the screen, but the image would be upside down. This proved Al-Haytham's theory that light travels in straight lines. It proved that the light from the bottom of the object outside of the pinhole would be projected to the top of the screen and vice-versa. The significance of the camera obscura was that it gave us a proper understanding of how our eyes work. [lxvii]

973 C.E.

Al-Biruni

First Anthropologist

Born Abu Rayhan Muhammad ibn Ahmad al-Biruni in 973 C.E. in modern day Uzbekistan. Al-Biruni was called the "Founder of Indology," "Father of Comparative Religion," "Father of Geodesy," and the first anthropologist. Al-Biruni was educated in the fields of physics, mathematics, astronomy, and natural sciences. He distinguished himself as a historian, chronologist, linguist, and although he did not identify himself as an anthropologist, he would have this recognition after his death. His linguistic abilities included, in addition to Arabic, Khwarezmian, Persian, Sanskrit, Greek, Hebrew and Syriac. Travel was a large part of his life where he devoted much of his lifetime to India in which he wrote a treatise called *Tarikh al-Hind* (History of India). He died at the age of 77 in modern day Afghanistan.

975 CE

Al-Mu'izz li-Din Allah
Al-Azhar University

Al-Azhar University is one of the most important universities and was founded in October of 975, in Cairo, Egypt by Al-Mu'izz li-Din Allah. The name Al-Azhar may have something to do with Fatimah, the prophets daughter, which is known as "al-zahra." The educational curriculum remained the same throughout the years, and the studies that it continues to provide are, Islamic law, theology, and Arabic language. When Egypt was conquered al-Azhar fell into an eclipse. In a later earthquake Al-Azhar had to be rebuilt. Al-Azhar is still standing today.
lxviii / lxix

980 CE
Bin-Sina (Avicenna)
Physician

Abū-ʿAlī al-Ḥusayn Bin-ʿAbdallāh Bin-Sīnā, born in August of 980 CE in Afshona, Uzbekistan was a very well-known Persian physician, philosopher, and scientist. He is credited with doing many remarkable things during his lifetime. He refined the scientific method, and wrote 450 books about astronomy, chemistry, geology, religion, logic, mathematics, physics, and poetry. His most famous book was called *Al-Shifa* (The Cure.) Some modern scholars have referred to him as the greatest thinker of the Golden Age of Islam.[lxx]

Bin-Sīnā, latinized as Avicenna, translated many of his idol, Aristotle's teachings, and transcripts. These translations helped influence European thought during the age of Enlightenment. He was acknowledged in the Islamic world as: "The Preeminent Master" (al-shaykh al-raʾīs), after Aristotle, whom Avicenna called "The First Teacher" (al-muʿallim al-awwal). Avicenna's most famous books are the; *The Book of Healing*, and *The Canon of Medicine* which remained an original medical text at many medieval universities and scholars for nearly 700 years. In 1973, Avicenna's *Canon of Medicine* was reprinted, and it made a resurgence in New York.[lxxi] / [lxxii]

Bin-Sina loved his religion and spent considerable time making attempts to connect Islam and philosophy. He later wrote commentaries on Quranic verses that relate to philosophy. Bin Sina also wrote many more books two of which he wrote under Al-Farabi. *"Compendium on the Soul"* about a Samanid ruler and "Philosophy for a Prosodist" about philosophy. He died June 22, 1037, in Hamedan, Iran. [lxxiii] / [lxxiv]

987 CE

Prince Abd Al-Rahman

Grand Mosque of Cordoba

The Grand Mosque of Cordoba was built in 987 CE in Cordoba, Spain. It is one of the oldest forms of construction still standing from the Muslim rule of Al-Andalus from the late eighth century and took nearly two hundred years to be completed. According to some historians' belief, the mosque was originally a temple for the Roman god, Janus. It was then converted to a church when the Visigoths gained control of Cordoba. Finally, it was changed into a mosque then fully rebuilt when the descendants of those who were exiled out of Umayya came. The prince encouraged building lavish buildings and agriculture, as well as decorating the infamous Mosque.

The building is decorated with a hypostyle hall filled with columns, a minaret which is used to call for prayer, a walkway that circles the courtyard, and an orange grove. The structure was made from recycled Roman parts that create an arch with gorgeous red bricks. The mosque is also known for the mihrab, which is the point that shows Muslims which way Mecca is to guide them in their prayer. The mihrab in this mosque is decorated in gold and has a luxurious arch as a frame. Another unique piece in the mosque is the horseshoe arch. This was a common piece for the Visigoths, and so, they placed it in their Church. It was then recycled by the Umayyads who rebuilt the church . And, as pictured above, there is the breathtaking dome. The dome is built with ribs that constantly cross each other, creating arches and finished off with gold.[lxxv]

988 CE
Ali Bin Ridwan
Supernova

Ali Bin Ridwan, born in Giza in 988 was an Egyptian physician, astronomer, and astrologer. He observed a supernova (now referred to as SN-1006) in 1006 C.E, of which he wrote significant commentaries. The supernova has been recorded as the brightest observed supernova in history. Scientists believe that the brightness was 16 times that of Venus. He also wrote commentaries on Ptolemy and Greek medicine. Even today, scientists still use his books as reference tools. He died in 1061.[lxxvi]

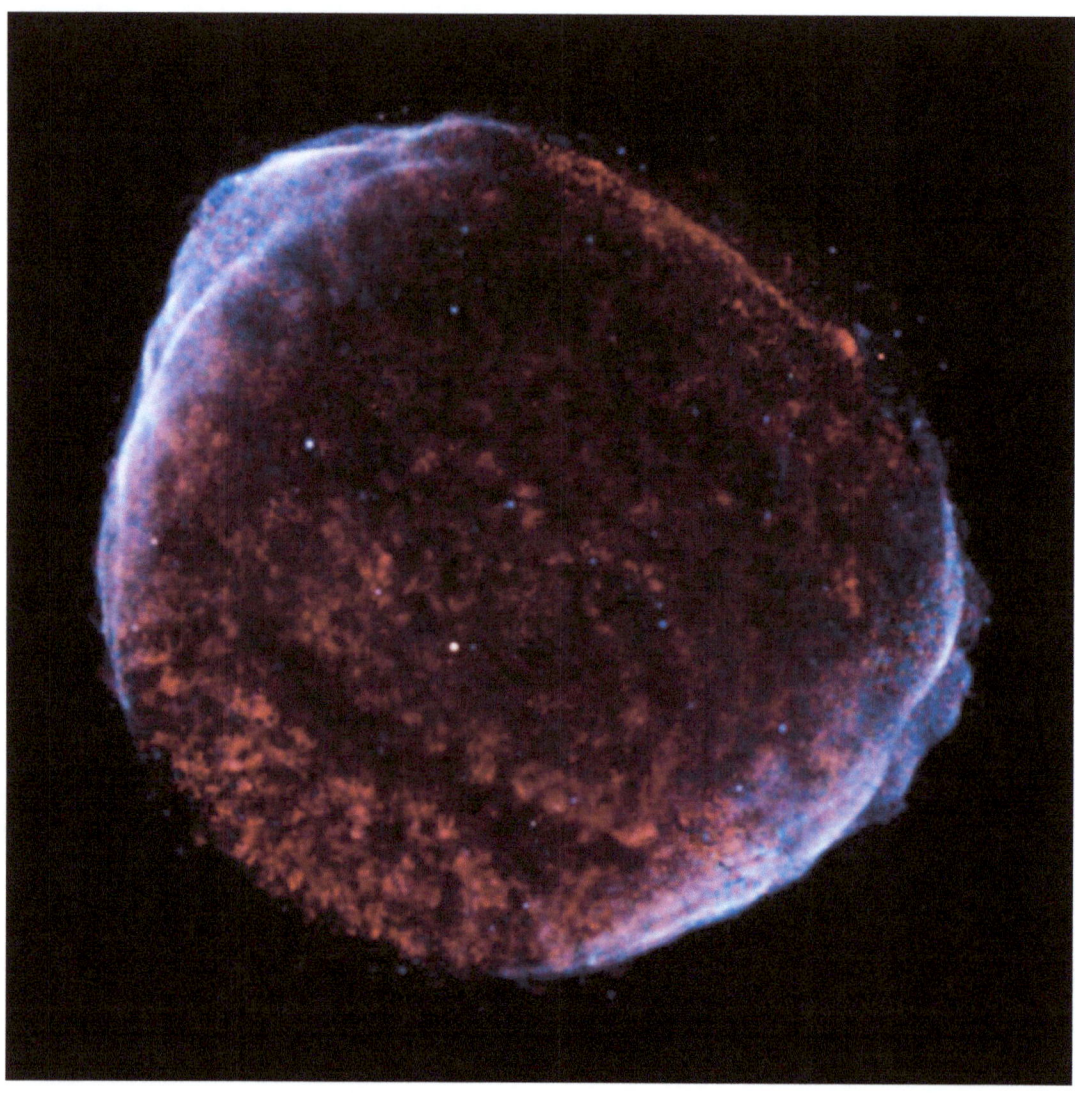

1058 CE

Abu Hamid Al Ghazali

Theologian

Al Ghazali was a great Muslim theologian. His full name was Abu Hamid Muhammad Bin Muhammad At Tusi Al Ghazali. Al Ghazali, despite travelling to various places throughout his life, still managed to die at his birthplace. Al Ghazali was born in 1058, and died in 1111, in Tusi, Iran. He was highly intelligent, and historians of the day recorded this. During the year 1091, Al Ghazali was asked to be the chief professor of the Nizamiyah collage in Baghdad. Al Ghazali took the opportunity, and while he was teaching, he was also commenting on the philosophies of Bin Sina, and Al Farabi.

Al Ghazali was a prolific writer and authored more than fifty works during his lifetime. His most famous work is called *The Revival of the Religious Sciences*. Al Ghazali also did philosophical studies in which he proved the claims of the Qur'an to be right, from challenges by other philosophers. Then he wrote *Maqasid Al Falasifah* (*The Aims of the Philosophers*).[lxxvii]

1080 CE
Abu Bakr
Philosopher

Abu Bakr Muhammad Bin Yahia Bin al-Sayigh (1080-1138} was born in Saragossa, Spain, and died in Fez, Morocco. He was the first notable philosopher and physician of Muslim Spain and wrote paraphrases of Aristotle's *Physics, Meteorology,* parts of *On Generation and Corruption,* and the spurious *De Plantis,* as well as extensive glosses on the logic of al-Farabi . His initial works include *Conduct of the Solitary,* a political work that deals with the predicament of the philosopher and his prevented desires in an imperfect state, and a short work titled *Conjunction with the Active Intellect,* which dealt with a favorite theme of Muslim Neoplatonists.[lxxviii]

1091 CE
Bin Zuhr
Cardiologist and Nutritionist

Abū Marwān ʿAbd al-Malik Bin Abī al-ʿAlāʾ Zuhr was also called Avenzoar or Abumeron. He was born in 1091, Sevilla Spain and died in 1162, also in Sevilla. Bin Zuhr disliked medical speculation, and for this reason he went against the teachings of the Persian master physician Avicenna. In his *Practical Manual of Treatments and Diet*, he described serious pericarditis (inflammation of the membranous sac surrounding the heart) and mediastinal abscesses (affecting the organs and tissues in the thoracic cavity above the diaphragm, excluding the lungs) and outlined surgical procedures for tracheotomy, excision of cataracts, and removal of kidney stones. He also discussed excessive contraction and dilation of the pupil (miosis and mydriasis) and promoted the use of the narcotic plant *Mandragora* as a treatment for ocular disease. His ("Practical Manual of Treatments and Diet") was later translated into Hebrew and Latin. In his time he was one of medieval Islam's foremost thinkers and the greatest medical clinician of the western caliphate. [lxxix]

1100 CE
Muhammad Al-Idrisi
Cartographer

Al-Idrisi was an Arab Muslim geographer, cartographer, and Egyptologist. He was born 1100 in Ceuta Spain and lived in Palermo, Sicily where he died in 1165. Al-Idrisi did lots of traveling in his time especially in his early life through Europe, Africa, and Asia and would map them as he did. He made corrections to the maps at his time that turned to be inaccurate. He studied at the University of Cordoba for many years. In 1154, he was able to create a map of the entire world with the information he acquired throughout his travels. [lxxx]

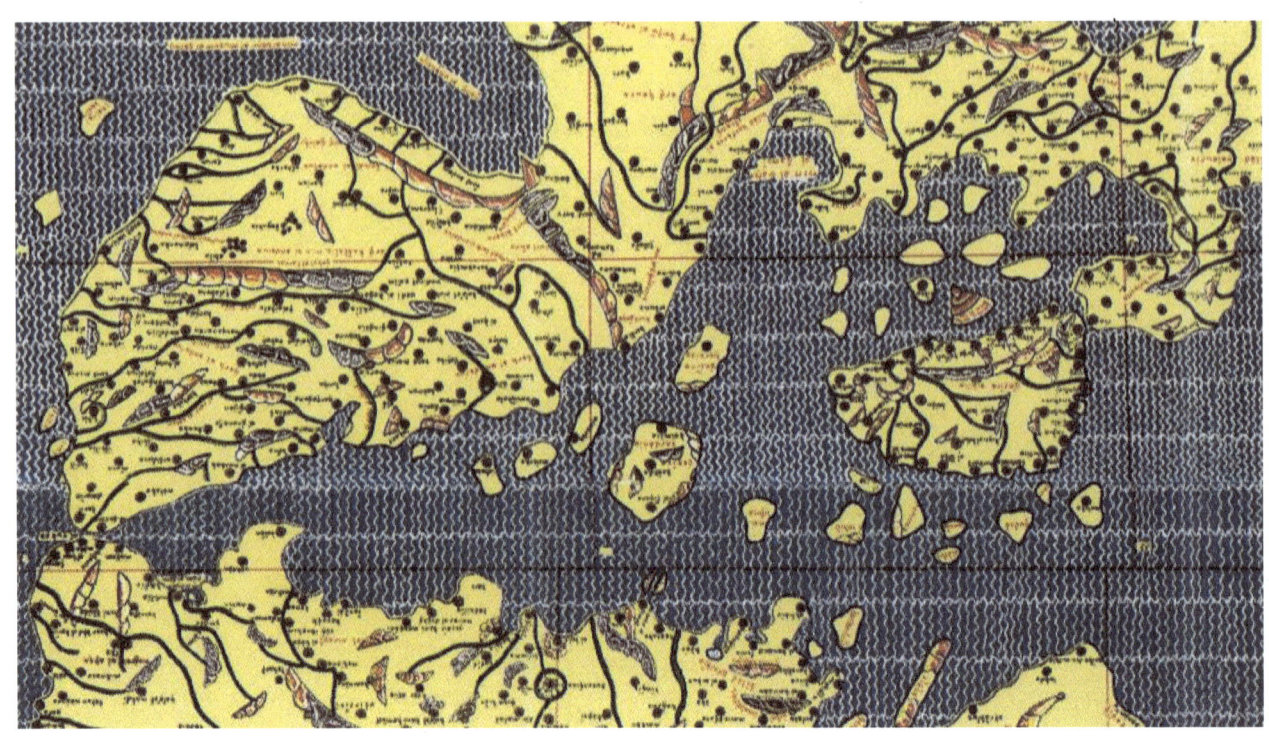

1126 CE
Bin Rushd (Averroes)
Polymath

Bin Rushd was born on April 14, 1126, and died December 10, 1198. He was a Muslim polymath worked with Aristotelian and Islamic philosophy. Both Bin Rushd and the scholar Bin Sina played a significant role in saving the works of Aristotle. In the 13th-century philosophical movement based on Averroes. Bin Rushd was born in Cordoba, in Spain. He died in Marrakech, Morocco.[lxxxi] / [lxxxii]

1136 CE
Al-Jazari
Mechanical Inventions

Ismail Al-Jazari was a Muslim scholar who lived during the golden age of Islam. Al-Jazari was born in the city of Jazirat bin Umar in 1136. It was from the name of the city that he is referred to simply as Al-Jazari.

He had many talents like being a renowned scholar, a skilled inventor, an artist, a craftsman, and an engineer. He designed more than 50 distinct types of devices including clocks, fountains, hand washing devices, musical devices, machines for raising water etc. He also wrote a booked called "The Book of Knowledge of Ingenious Mechanical Devices" and was written in 1206. He died later that year. [lxxxiii] / [lxxxiv]

Al-Jazari was a Turkish artisan, inventor, scholar, mathematician, and engineer. His treatise *The Book of Knowledge of Ingenious Mechanical Devices*, wrote about one hundred mechanical devices that Al-Jazari invented and further how to build them. It was among the first books to have the "do it yourself" philosophy, as he was less interested in the theory or technology behind making an object, and more interested in the actual process of making it. He invented many objects and devices, including the crankshaft and the first automatic doors and gates. Al-Jazari was significant because he paved the way for many engineers and craftsmen to come after him, including Leonardo Da Vinci. He was also an important figure in engineering because he created some of the first robots and automata. His most famous invention was the Elephant Clock.

The Elephant Clock

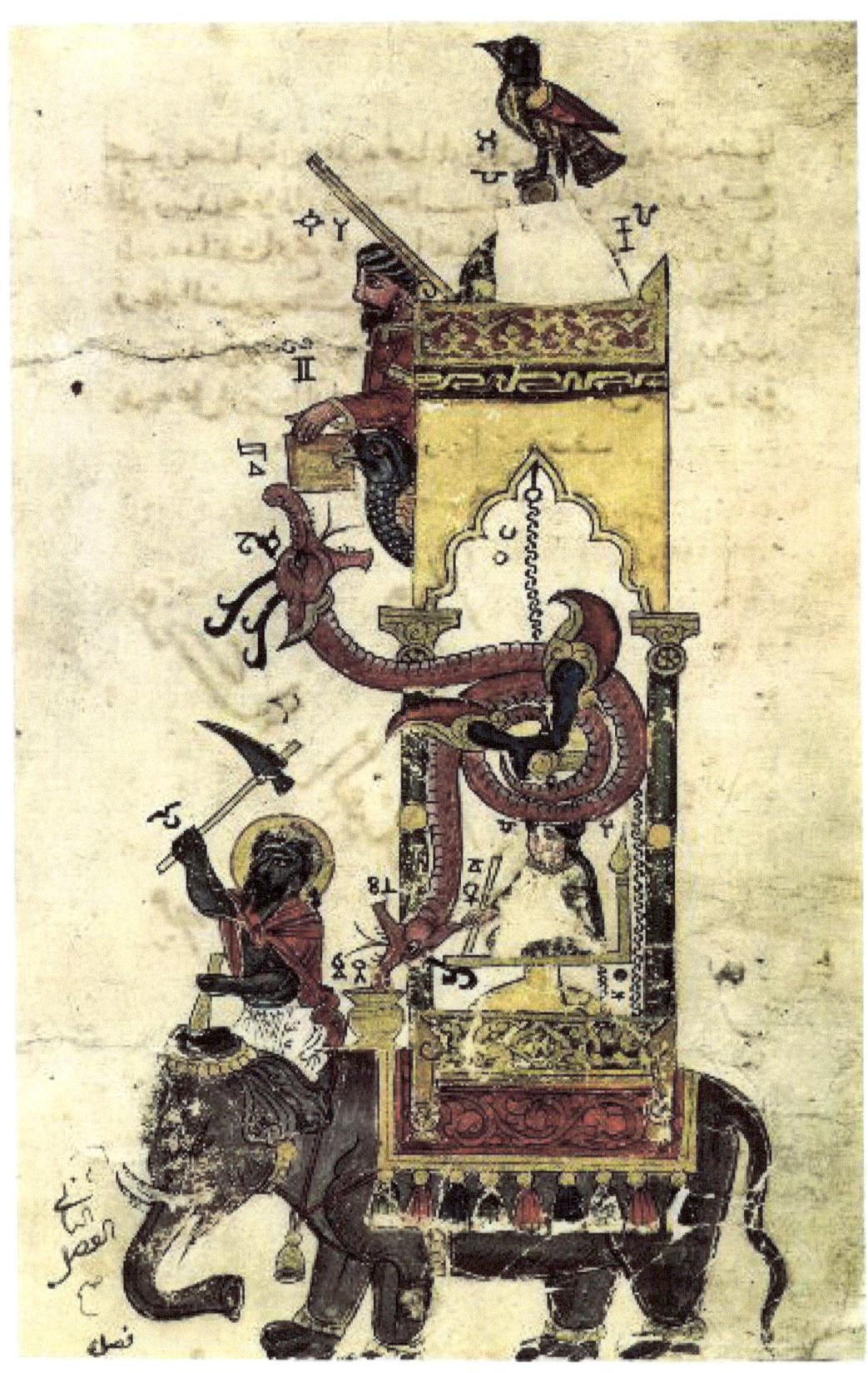

The Elephant Clock operated with a timing device, hidden inside the elephant, and based on a water-filled basin. In the basin, there is a bucket, and inside the bucket, there is a deep bowl which floats in the water with a small hole in its center. After half an hour, the bowl fills with water entering through the center hole, and it pulls a string attached to the top tower on the elephant. The see-saw mechanism then releases a ball that falls into a serpent's mouth and from the weight it leans forward and pulls the bowl out of the water. When this action starts, the other figures in the top tower start to move. There is a shape that, raises its hands when the serpent leans forward, while the driver of the elephant is hitting a drum. The serpent then goes back to its original position. This cycle can last between 30 minutes to an hour depending on the bowl used. The cycle repeats itself. It is important that in the upper container there are enough balls that can power the emptying of the bowl. The Elephant Clock represents diverse cultures: the elephant is from India, the serpents are from China, and the turban is from Islamic culture.[lxxxv]

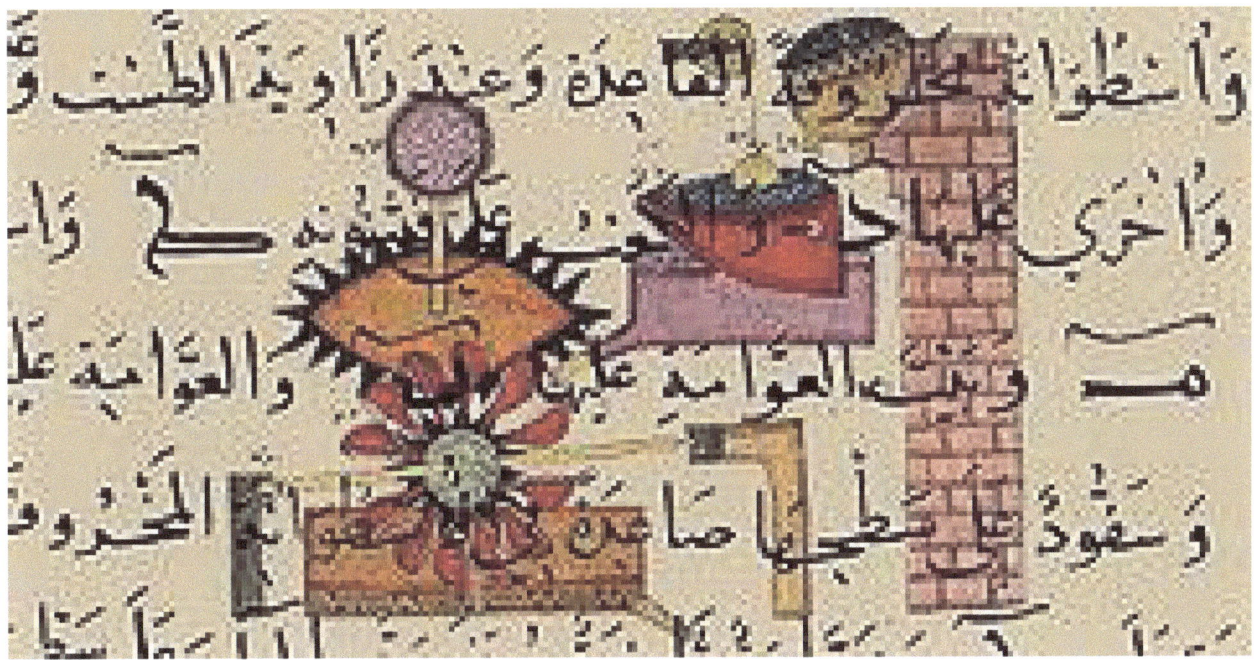

1137 CE

Yusuf ibn Ayyub ibn Shadi
Saladin
Sultan of Egypt and Syria

Saladin was born 1137 in Tikrit, Iran, and died from a fever on March 4, 1193. Saladin studied multiple things like Islam, Mathematics, Philosophy, and Law. When Saladin was fourteen, he started to work in the military with his uncle. Saladin and his uncle started to gain power in Egypt, they went to fight the crusades in Europe, and the fight was remarkably successful. When Saladin's uncle died (at the age of 31) he was supposed to be the leader of the Syrian troops. Salafin defeated the crusader army and went to Jerusalem. He defeated the city and became remarkably successful. After the fight between the crusaders and Jerusalem, Christians in Europe had started a third crusade, Saladin's troops had horrible defeats. In 1192 Saladin and King Richard had signed a treaty, and it just said that Jerusalem was kept in the Muslims responsibility, but they were allowed for the Christian pilgrims. Shortly after signing the treaty Saladin died from a fever. [lxxxvi] / [lxxxvii]

1149 CE
Fakhr al-Din al-Razi
Theologian

Fakhr al din al Razi was an Iranian Theologian and philosopher. He was born in Rayy in 1149 and died in Herat (modern day Afghanistan) in 1209. He wrote *Tafsir Al-Alkabeer* on the Quran which means The Great Commentary. He also authored other books about logic and medicine in addition to other topics. [lxxxviii] He was inspired as a youth to study philosophy and later theology. [lxxxix] / [xc]

1154 CE
Sohrevardi
Philosopher of Illusionism

Sohrevardi (1154-1191) was a Persian Philosopher otherwise known as Shihab al-din. He was born in Sohrevard, Iran in 1154 CE. He died in 1191 CE in Aleppo, Syria. He wrote dozens of books about philosophy and mysticism. Sohrevardi even built an Islamic school based on platonic ideas. He became the founder of the school of illusionism.[xci]

1191 CE
Melike Mama Hatun
Turkish Bath

It does not exactly say when and where Melike Mama Hatun was born and when and where she died, but it does say that she lived during the 12th century. Melike Mama Hatun was a ruler of the Saltukid dynasty for about nine years, from 1191-1200. When Melike Mama Hatun was still ruling she had a caravanserai, which is an inn for travelers to rest in; she also had a mosque, a bridge, and a hammam , which is also known as a Turkish steam bath. All of these were built in the town of Tercan where they remain, all renamed in her honor. [xcii]

1200 CE
Mo'ayeduddin Urdi
Model of Planetary Motion

Mo'ayeduddin Urdi, born in 1200 was an Arab astronomer, architect, mathematician, and engineer. He was born in present day Syria and then moved to Persia. He is well known for developing a model of planetary motion. He was responsible in Syria for constructing a water supply installation. He died in 1266 [xciii]

1201 CE
Nasir Al din Al Tusi
Rational Science

Nasīr al-Dīn al-Tūsī was born on 18 February 1201 in Tus, Khorasan (now Iran) and he died on 26 June 1274 in Kadhimain (near Baghdad now in Iraq). He was a Muslim Persian scholar and creative writer in various parts of science and philosophy. He was also an astronomer, mathematician, physicist, philosopher, and theologian. Al-Tūsī was born into a family of scholars. His father and his uncles wanted him to pursue the Islamic religious sciences and the rational sciences. Rational sciences were defined as the compliance of one's belief and why they believe that belief. He studied philosophy and mathematics in his hometown Tūs, but eventually had to travel to Nīshāpūr (after 1213) to continue his education in sciences, medicine, and philosophy.

He studied the works of Bin Sīnā, who became an important developing influence. Al-Tūsī then traveled to Iraq where his he studied legal theory; in Mosul (sometime between 1223 and 1232), one of his teachers was Kamāl al-Dīn Bin Yūnus (died 1242), a legal scholar who was also known for his knowledge in astronomy and mathematics. Al-Tūsī also wrote over 150 works, in Arabic and Persian, which talked about mathematical sciences, philosophy, and the religious sciences (*fiqh, kalām,* and Sufism). The wide diffusion of his works and their influence resulted in his titles *khwāja* (distinguished scholar and teacher), *ustādh al-bashar* (teacher of mankind), and *al-mu'allim al-thālith* (the third teacher, after Aristotle and Al-Fārābī). Al-Tūsī was the director of the Islamic major astronomical center of Marāgha (northern Iran). [xciv]

1213 CE

Bin Al-Nafis

Physician

Bin al Nafis, born in 1213 was and Arab physician from Damascus but moved to Cairo for his work. Bin al Nafis has made huge contributions to the understanding of pulmonary circulation. He also authored books in many fields such as medicine, law, philosophy and many more fields. One of his most famous writing Bin al Nafis did was the *"Commentary on Anatomy in Avicenna's Canon."* Another famous writing of his was the *"Comprehensive Book of Medicine"* even though this encyclopedia was not completed, it is still useful for many scholars. Bin al Nafis made many important contributions and will always be remembered for them. He died in 1288. [xcv]

1236 CE
Yahya Bin Mahmud al-Wasiti
Maqamat of Al-Hariri

Visual art was prominent throughout the entirety of the Golden Age of Islam, coming in the form of painting, ceramics, calligraphy, architecture, textiles, and metalwork. A prominent painter and calligrapher from the Golden Age of Islam was Yahya Bin Mahmud al-Wasiti, known for producing ninety-six illustrations in the *Maqamat of Al-Hariri*, a series of anecdotes depicting Islamic life in the thirteenth century. Tin-opacified glazing was an innovative technology developed in the Golden Age by Islamic potters, along with many other innovations brought to ceramics by the Golden Age of Islam. Calligraphy was ubiquitous during this period, and was used in architecture, paintings, sculptures, pottery, and coins. Its significance during the Golden Age was that it was so strongly connected to the Qur'an because this is where most Islamic calligraphy is based. Architecture played a considerable role in Islamic art because it encompasses several elements of Islamic art, including calligraphy and tilework. Islamic architecture uses domes, columns, and arches, influenced by Roman, Byzantine, Persian, Mesopotamian, and all other areas the Muslims conquered during the eighth and ninth centuries. Textiles of the Golden Age of Islam included carpets, rugs, and embroidery, and reflected a similar style to the architecture of that time. This is how visual art played a role in the Golden Age of Islam. [xcvi / xcvii / xcviii / xcix / c / ci / cii / ciii]

1237 CE
Spinning Wheel

The spinning wheel was invented in the Islamic world by the early 11th century. There is evidence pointing to the spinning wheel being known in the Islamic world by 1030, and the earliest clear illustration of the spinning wheel is from Baghdad, drawn in 1237.[97]

Madrasa

The madrasa was basically the first college in the world. Madrasas were established to create a higher level of learning. In the Golden Age of Islam, Madrasa referred to a religious school dedicated to Islam. The madrasa would teach Islamic theology, Qur'an, Hadith, literature, mathematics, astronomy, natural sciences, and law. The madrasa's curriculum consisted of professors giving lectures and textbook memorization. The colleges were free, and so was the food and health care. Royal families would often donate money to help the schools get richer in terms of facility, and curriculum. The impact of madrasas to the western Arabian world, and North Africa were advances in architecture, and astronomy. Madrasas were one of the main reasoning behind all the astronomical inventions made by Arabian astronomers. Overall, the madrasa benefited the Arabian world in terms of economy, inventions, and advances. [civ]

BIBLIOGRAPHY

[i] *"Battle of the Zab." World History Project, worldhistoryproject.org/750/1/25/battle-of-the-zab.*
[ii] Canfield, Robert L. (2002). Turko-Persia in Historical Perspective. Cambridge University Press. p. 5. ISBN 9780521522915.
[iii] "ABŪ MOSLEM ḴORĀSĀNĪ – Encyclopaedia Iranica". www.iranicaonline.org. Archived from the original on 22 November 2015. Retrieved 20 November 2015.
[iv] Finer, S. E. (1 January 1999). The History of Government from the Earliest Times: Volume II: The Intermediate Ages p.720. OUP Oxford. ISBN 9780198207900.
[v] Holt, Peter M. (1984). "Some Observations on the 'Abbāsid Caliphate of Cairo". Bulletin of the School of Oriental and African Studies. University of London. 47 (3): 501–507. doi:10.1017/s0041977x00113710
[vi] https://en.wikipedia.org/wiki/As-Saffah
[vii] "Name-New world encyclopedia. Author-Jafar Al-Sadiq.Date-March 30, 2014 (http://www.newworldencyclopedia.org/entry/Jabir_Bin_Hayyan)
[viii] http://muslimheritage.com/article/beginning-paper-industry

[ix] *Hawting, G.R. "Al-Manṣūr." Encyclopædia Britannica, Encyclopædia Britannica, Inc., www.britannica.com/biography/al-Mansur-Abbasid-caliph*
[x] Huda. "The Role Baghdad Has Played in Islamic History." ThoughtCo, ThoughtCo, 12 Sept. 2018, www.thoughtco.com/baghdad-in-islamic-history-2003773.
[xi] Charette , Francois. "Ḥabash Al-Ḥāsib: Abū Jaʿfar Aḥmad Bin ʿAbd Allāh Al-Marwazī." *Islamic Science*, islamsci.mcgill.ca/RASI/BEA/Habash_al-Hasib_BEA.htm
[xii] "'Habash Al-Hasib Al-Marwazi' ." *Revolvy*, www.revolvy.com/page/Habash-al%252DHasib-al%252DMarwazi.

[xiii] Ancient-Origins. "The House of Wisdom: One of the Greatest Libraries in History." *Ancient Origins*, Ancient Origins, 1 Jan. 2017, www.ancient-origins.net/``eancient-places-asia/house-wisdom-one-greatest-libraries-history-007292.

[xiv] "The House of Wisdom: Baghdad's Intellectual Powerhouse | 1001 Inventions." 1001Inventions, 1001inventions.com/house-of-wisdom.
[xv] Britannica, The Editors of Encyclopaedia. "Al-Khwārizmī." *Encyclopædia Britannica*, Encyclopædia Britannica, Inc., 17 Feb. 2017, https://www.britannica.com/biography/al-Khwarizmi.
[xvixvi] "Home." *Famous Scientists*, www.famousscientists.org/muhammad-Bin-musa-al-khwarizmi/.
[xvii] "Muhammad Bin Musa Al-Khwarizmi." Famous Scientists, www.famousscientists.org/muhammad-Bin-musa-al-khwarizmi/
[xviii] Bourne, Murray. "Al-Khwarizmi, the Father of Algebra." *Intmathcom RSS*,

www.intmath.com/basic-algebra/al-khwarizmi-father-algebra.php.

[xix] Islamic Mathematics - The Story of Mathematics, www.storyofmathematics.com/islamic.html.
[xx] Sourdel, Dominique. "Al-Maʾmūn." Encyclopædia Britannica, Inc., 1 Jan. 2019, www.britannica.com/biography/al-Mamun.
[xxi] Tech@whyislam.org. "Fatima Al-Fihri: Founder of World's Very First University." *WHY*, www.whyislam.org/muslim-heritage/fatima-al-fihri-founder-of-worlds-very-first-university/.
[xxii] Tech@whyislam.org. "Fatima Al-Fihri: Founder of World's Very First University." WHY, www.whyislam.org/islam/fatima-al-fihri-founder-of-worlds-very-first-university/.
[xxiii] *"Extraordinary Women from the Golden Age of Muslim Civilisation | 1001 Inventions." 1001Inventions, www.1001inventions.com/womensday*
[xxiv] *Tech@whyislam.org. "Fatima Al-Fihri: Founder of World's Very First University." WHY,www.whyislam.org/muslim-heritage/fatima-al-fihri-founder-of-worlds-very-first-university/.*
[xxv] http://www.manchesteruniversitypress.co.uk/articles/fatima-al-fihri-founder-worlds-firstuniversity/

[xxvi] *"Al-Kindi." Al-Kindi | Muslim Heritage, www.muslimheritage.com/article/al-kindi*
[xxvii] "Abu Yusuf Yaqub Bin Ishaq Al-Sabbah Al-Kindi." Al-Kindi (about 805-873), www-groups.dcs.st-and.ac.uk/history/Biographies/Al-Kindi.html.
[xxviii] "Abu Yusuf Yaqub Bin Ishaq Al-Sabbah Al-Kindi," www-groups.dcs.st-and.ac.uk/history/Biographies/Al-Kindi.html.
[xxix] "Al-Kindi." FAMOUS INVENTORS, www.famousinventors.org/al-kindi.
[xxx] *Casulleras, Josep (2007).* "Banū Mūsā". *In Thomas Hockey; et al. (eds.). The Biographical Encyclopedia of Astronomers. New York: Springer. pp. 92–4.* ISBN 978-0-387-31022-0. *(PDF version*

[xxxi] "Hunayn Bin Ishaq Al-'Ibadi, Abu Zayd.". "Hunayn Bin Ishaq Al-'Ibadi, Abu Zayd." Complete Dictionary of Scientific Biography, Encyclopedia.com, www.encyclopedia.com/science/dictionaries-thesauruses-pictures-and-press-releases/hunayn-Bin-ishaq-al-ibadi-abu-zayd.

[xxxii] "Abu Zayd Hunayn Bin Ishaq al-Ibadi." Hunayn Bin Ishaq (808-873). <http://www-history.mcs.st-andrews.ac.uk/Biographies/Hunayn.html>.

[xxxiii] Britannica, The Editors of Encyclopaedia. "Aḥmad Bin Ṭūlūn." Encyclopædia Britannica, Encyclopædia Britannica, Inc., 25 Feb. 2019, www.britannica.com/biography/Ahmad-Bin-Tulun.

[xxxiv] "Thabit Bin Qurra ." Famous Scientists, www.famousscientists.org/thabit-Bin-qurra/.

[xxxv] "Home." Famous Scientists, www.famousscientists.org/al-battani/.

[xxxvi] Britannica, The Editors of Encyclopaedia. "Al-Rāzī." Encyclopædia Britannica, Encyclopædia Britannica, Inc., 29 Mar. 2018, www.britannica.com/biography/al-Razi.

[xxxvii] "Islamic Culture and the Medical Arts: Al-Razi, the Clinician." *U.S. National Library of Medicine*, National Institutes of Health, 15 Dec. 2011, www.nlm.nih.gov/exhibition/islamic_medical/islamic_06.html.

[xxxviii] "Name-Nationwidefuels.Author-nationwidefuels.Cite-(https://www.nationwidefuels.co.uk/9-things-didnt-know-kerosene/)

[xxxix] The editors of Wikipedia "Soap." Wikipedia, Wikimedia Foundation, en.wikipedia.org/wiki/Soap.

[xl] Hendricks, Scotty. "10 Golden Age Philosophers, and Why You Should Know Them." Big Think, Big Think, bigthink.com/scotty-hendricks/ten-islamic-philosophers-you-dont-know-and-why-you-shoud.

[xlixli] "Abu Nasr Al-Farabi." Famous Scientists, www.famousscientists.org/abu-nasr-al-farabi/.

[xlii] "An Introduction and Biography of Al-Farabi." The Great Thinkers, The Foundation of Constitutional Government Inc., thegreatthinkers.org/al-farabi/introduction/.

[xliii] "Name-Forgotten Islamic History.Author-Naeem Ali.Date-November 3rd, 2013. Cite-(http://www.forgottenislamichistory.com/2013/11/abbas-Bin-firnas-worlds-first-pilot.html)

[xliv] "Saadia Gaon סעדיה גאון." Sefaria, www.sefaria.org/person/Saadia%20Gaon.

[xlv] Pessin, Sarah. "Saadya [Saadiah]." Stanford Encyclopedia of Philosophy. 06 May 2003. Stanford University. <https://plato.stanford.edu/entries/saadya/>.

[xlvi] Zucker, Moses. "Saʿadia ben Joseph." Encyclopædia Britannica. 01 Jan. 2019. Encyclopædia Britannica, inc. <https://www.britannica.com/biography/Saadia-ben-Joseph>.

[xlvii] Bin 'Adi, Yahya (893-974), www.muslimphilosophy.com/ip/rep/H034

[xlviii] al-Masudi.", "Ali Bin al-Husayn. "Ali Bin Al-Husayn Al-Masudi." Encyclopedia of World Biography, Encyclopedia.com, 2019, www.encyclopedia.com/history/encyclopedias-almanacs-transcripts-and-maps/ali-Bin-al-husayn-al-masudi.

[xlix] https://www.staff.science.uu.nl/~gent0113/alsufi/alsufi_biography.htm

[l] https://www.eso.org/gen-fac/pubs/astclim/espas/iran/sufi.html

[li] "Al-Hakim Nishapuri." Wikiwand, www.wikiwand.com/en/Al-Hakim_Nishapuri.

[lii] Hajar, Rachel. "The Air of History Part III: The Golden Age in Arab Islamic Medicine An Introduction." Heart Views : the Official Journal of the Gulf Heart Association, Medknow Publications & Media Pvt Ltd, 2013, www.ncbi.nlm.nih.gov/pmc/articles/PMC3621228/.

[liii] "Al-Hakim Nishapuri." Wikiwand, www.wikiwand.com/en/Al-Hakim_Nishapuri.

[liv] Boyle, John Andrew. "Ferdowsī." Encyclopædia Britannica, Encyclopædia Britannica, Inc., 6 Feb. 2019, www.britannica.com/biography/Ferdowsi.

[lv] "Name-Muslim Heritage. Author-Ibrahim Shaikh. Cite- (http://muslimheritage.com/article/abu-al-qasim-al-zahrawi-great-surgeon)

[lvi] "Name-Reality is often Bitter.Author-ftikhar Ajmal Bhopal. Date-December 14, 2009.(Cite-https://iabhopal.wordpress.com/2009/12/14/who-invented-surgical-instruments-2/)

[lvii] "Abu Al-Wafa Al-Buzjanî." *Abu Al-Wafa Al-Buzjanî | Muslim Heritage*,muslimheritage.com/article/abu-al-wafa-al-buzjan%C3%AE.

[lviii] "Name-Mariam Al-Astulabiya-Great Muslim minds|CABTV. Author-CABTV.Date-September 13, 2018.Cite-(https://www.youtube.com/watch?v=UHGlfS2MX2c)

[lix] "Tech Women: Meet Mariam Astrulabi , the Woman behind Astrolabes." SheThePeople TV, 2 Dec. 2017, www.shethepeople.tv/news/tech-women-meet-mariam-astrulabi-the-woman-behind-astrolabes.

[lx] https://medium.com/@travelingwritvr/astrolabe-during-the-time-of-islamic-golden-ages-49504641dfda

[lxi] "Name-Smithsonian.com,THINK BIG.Author-Smithsonian.Cite-(https://www.smithsonianmag.com/innovation/astrolabe-original-smartphone-180961981/)

[lxii] "Al-Muqaddasi: The Geographer from Palestine." *Al-Muqaddasi: The Geographer from Palestine | Muslim Heritage*, http://muslimheritage.com/article/al-muqaddasi-geographer-palestine.

[lxiii] "Name-1001 Inventions and the World of Bin Al-Haytham.Author not listed. Cite-(http://www.Binalhaytham.com/discover/who-was-Bin-al-haytham/)

[lxiv] "Abu Ali Al-Hasan Bin Al-Haytham." www-history.mcs.st-andrews.ac.uk/Biographies/Al-Haytham.html.

[lxv] "The Reason That Your Phone Has a Camera." The.Ismaili, 5 July 2017, the.ismaili/our-culture/reason-your-phone-has-camera. "Who Was Bin Al-Haytham - Bin Al." Haytham, www.Binalhaytham.com/discover/who-was-Bin-al-haytham/.

[lxvi] Elsevier. "How an Ancient Muslim Scientist Cast His Light into the 21st Century." Elsevier Connect, www.elsevier.com/connect/how-an-ancient-muslim-scientist-cast-his-light-into-the-21st-century.

[lxvii] 1001Inventions. "Bin Al-Haytham - The First Scientist - Origin of the Camera." YouTube, YouTube, 5 Nov. 2009, www.youtube.com/watch?v=a5icY1dMin4.

[lxviii] "Al-Azhar University." Al-Azhar University | Muslim Heritage. <http://muslimheritage.com/article/al-azhar-university>.

[lxix] "Al-Azhar University." Encyclopædia Britannica. 17 Oct. 2017. Encyclopædia Britannica, inc. <https://www.britannica.com/topic/al-Azhar-University>.

[lxx] "Eval(ez_write_tag([[970,90],'newworldencyclopedia_org-Box-2','ezslot_0',106,'0']));Avicenna." Avicenna - New World Encyclopedia, www.newworldencyclopedia.org/entry/Avi

[lxxi] Gutas, Dimitri. "Bin Sina [Avicenna]." *Stanford Encyclopedia of Philosophy*, Stanford University, plato.stanford.edu/entries/Bin-sina/.

[lxxii] Hendricks, Scotty. "10 Golden Age Philosophers, and Why You Should Know Them." *Big Think*, Big Think bigthink.com/scotty-hendricks/ten-islamic-philosophers-you-dont-know-and-why-you-shoud.

[lxxiii] "Philosophy for a Prosodist" about philosophy. Internet Encyclopedia of Philosophy, www.iep.utm.edu/avicenna/.

[lxxiv] "Bin Sina's The Canon of Medicine." Bin Sina's The Canon. of.Medicine | Muslim Heritage, muslimheritage.com/article/Bin-sinas-canon-medicine.

[lxxv] Mirmobiny , Shadieh. "The Great Mosque of Cordoba." Khan Academy, www.khanacademy.org/humanities/ap-art-history/early-europe-and-colonial-americas/ap-art-islamic-world-me dieval/a/the-great-mosque-of-cordoba.

[lxxvi] "'Three Times Greater than Venus': Bin Ridhwan's Observation of Supernova 1006." "Three Times Greater than Venus": Bin Ridhwan's Observation of Supernova 1006 | Muslim Heritage, http://muslimheritage.com/article/three-times-greater-venus-Bin-ridhwans-observation-supernova-1006.

[lxxvii] Watt, William Montgomery. "Al-Ghazālī." Encyclopædia Britannica, Encyclopædia Britannica, Inc., 7 Feb. 2019, www.britannica.com/biography/al-Ghazali.

[lxxviii] "Bin Bajjah." *Oxford Islamic Studies Online*, www.oxfordislamicstudies.com/article/opr/t125/e934.

[lxxix] Britannica, The Editors of Encyclopaedia. "Bin Zuhr." *Encyclopædia Britannica*, Encyclopædia Britannica, Inc., 1 Jan. 2019, www.britannica.com/biography/Bin-Zuhr.

[lxxx] "Who Was Muhammad Al-Idrisi? Everything You Need to Know." Muhammad Al-Idrisi Biography - Childhood, Life Achievements & Timeline, www.thefamouspeople.com/profiles/muhammad-al-idrisi-7511.php.

[lxxxi] Internet Encyclopedia of Philosophy, www.iep.utm.edu/Binrushd/.

"Abu'l Hasan Ahmad ibn Ibrahim al-Uqlidisi | Encyclopedia.com". www.encyclopedia.com.

^ Jump up to:[a][b] *O'Connor, John J.; Robertson, Edmund F., "Abu'l Hasan Ahmad ibn Ibrahim Al-Uqlidisi", MacTutor History of Mathematics archive, University of St Andrews*

^ Jump up to:[a][b] *Berggren, J. Lennart (2007). "Mathematics in Medieval Islam". The Mathematics of Egypt, Mesopotamia, China, India, and Islam: A Sourcebook. Princeton University Press. p. 518. ISBN 978-0-691-11485-9.*

The Arithmetics of Al-Uqlisidi, The story of Hindu-Arabic Arithmetic, translated and annotated by A.S.Saidan, D.Reidel Publishing Company,Boston, 1978 [lxxxii] McLean, John. "World Civilization." The Islamic Golden Age | World Civilization, courses.lumenlearning.com/suny-hccc-worldcivilization/chapter/the-islamic-golden-age/.

[lxxxiii] https://www.thefamouspeople.com/profiles/ismail-al-jazari-38170.php

[lxxxiv] https://www.famousinventors.org/al-jazari

[lxxxv] "The Elephant Clock: One of the Greatest Inventions of the Outstanding Mechanical Engineer Al-Jazari." The Vintage News, 6 May 2017, www.thevintagenews.com/2017/05/06/the-elephant-clock-one-of-the-greatest-inventions-of-the-outstanding-mechanical-engineer-al-jazari/.

[lxxxvi] "Early Islamic World: Biography." Ducksters Educational Site. <https://www.ducksters.com/history/islam/saladin.php>

[lxxxvii] Walker, Paul E. "Saladin." Encyclopædia Britannica. Encyclopædia Britannica, inc. <https://www.britannica.com/biography/Saladin>.

[lxxxviii] https://www.google.com/amp/s/bigthink.com/ten-islamic-philosophers-you-dont-know-and-why-you-shoud-2604502887.amp.html

[lxxxix] Al-Razi, Fakhr al-Din (1149-1209). <http://www.muslimphilosophy.com/ip/rep/H044.htm>.

[xc] Haywood, John A. "Fakhr ad-Dīn ar-Rāzī." Encyclopædia Britannica. 01 Jan. 2019. Encyclopædia Britannica, Inc. <https://www.britannica.com/biography/Fakhr-ad-Din-ar-Razi#ref234420>.

[xci] Philosophy of Mysticism, www.sunypress.edu/p-6244-philosophy-of-mysticism.aspx.

[xcii] Revolvy, LLC. ""Melike Mama Hatun" on Revolvy.com." Revolvy. <https://www.revolvy.com/page/Melike-Mama-Hatun>.

[xciii] Ilkhani, Zig-i. "Mo'ayyeduddin Urdi." Enacademic, http://enacademic.com/dic.nsf/enwiki/922903.

[xciv] "Nasir Al-Din Al-Tusi." *Nasir Al-Din Al-Tusi (1201-1274)*, www-history.mcs.st-andrews.ac.uk/Biographies/AlTusi_Nasir.html.

[xcv] "Bin Al-Nafis, the Pulmonary Circulation, and the Islamic Golden Age." Journal of Applied Physiology, www.physiology.org/doi/full/10.1152/japplphysiol.91171.2008

[xcvi] McLean, John. "World Civilization." *The Islamic Golden Age | World Civilization*, courses.lumenlearning.com/suny-hccc-worldcivilization/chapter/the-islamic-golden-age/.

[xcvii] Makiya, Kanan (1991). The Monument: Art, Vulgarity, and Responsibility in Iraq. University of California Press. ISBN 9780520073760.

[xcviii] Ankori, Gannit (2013). Palestinian Art. Reaktion Books. ISBN 9781780232416.

[xcix] Explorers: Tales of Endurance and Exploration. Penguin. 2010. ISBN 9780756675110.

[c] Jonathan Bloom and Sheila S. Blair (eds), Grove Encyclopedia of Islamic Art & Architecture, Oxford University Press, 2009, p.210; Kember, P. (ed.), Benezit Dictionary of Asian Artists, Oxford University Press, 2012, ISBN 9780199923014

[ci] Jonathan Bloom and Sheila S. Blair (eds), Grove Encyclopedia of Islamic Art & Architecture, Oxford University Press, 2009, p.294

[cii] Wijdan, A. (ed.), Contemporary Art From The Islamic World, p.166

[ciii] "Yaḥyā Bin Maḥmūd al-Wāsiṭī," in: Encyclopædia Britannica, Online:

[civ] Britannica, The Editors of Encyclopaedia. "Madrasah." Encyclopædia Britannica, Encyclopædia Britannica, Inc., 12 Feb. 2018, www.britannica.com/topic/madrasah